American Classics

CHEVROLET
In the Sixties

Robert Genat

Motorbooks International
Publishers & Wholesal

Dedication

To my big brother, Jim. Thanks for introducing me to the wonderful world of Chevys more than 35 years ago, and thanks for letting me drive your new 409 four-speed 1963 Impala to my senior prom. I guess I can tell you now—I drag raced it on Telegraph Road against a Pontiac. I won!

First published in 1997 by Motorbooks International Publishers & Wholesalers, 729 Prospect Avenue, PO Box 1, Osceola, WI 54020-0001

Library of Congress Cataloging-in-Publication Data Available

ISBN 0-7603-0209-X

On the front cover: Chevrolet pulled out all the stops for the new 1963 Corvette. The Sting Ray shown in this Chevrolet studio shot carried the 360-horsepower fuel-injected 327. In the intervening years, the Sting Ray body of 1963–1967 has become *the* classic Corvette shape. *GM Media Archives*

On the frontispiece: Two of the sixties greatest performance cars, a 1969 Z/28 Camaro and a 1966 big-block Corvette.

On the title page: Chevy had it all in the sixties, style, performance, luxury. From the family garage, to the boulevard, to the race track, Chevrolet covered all the bases.

On the back cover: Top: With a versatile line-up like this, it's no surprise that Chevrolet posted record sales of 2.3 million units in 1964. *GM Media Archives* Bottom: Corvairs roll off the assembly line in the happy pre-Nader days of 1960. *GM Media Archives*

Printed in Hong Kong

Contents

Acknowledgments

Like most men, my true relationship with automobiles started when I took driver's training. The first car I ever drove was a blue 1961 Chevy Bel Air four-door sedan. It had a six-cylinder engine, a three-speed manual transmission, blackwall tires, and small hubcaps. Chevrolet provided an optional passenger-side brake pedal for my high school instructor, Mr. George Fairall, so he could at least try to save his life in case I screwed up too badly. There are certain events one never forgets, and this is one of them.

My thanks to current and former GM/Chevrolet employees whose skill and inventiveness shaped the cars in this book. I especially want to thank the following former Chevrolet employees for their honesty and candor during my lengthy interviews with them: Chuck Jordan, Don Schwarz, Dave Holls, Dick Keinath, Blaine Jenkins, Alvie Smith, Larry Shinoda, Al Flowers, and Tom Semple. Thanks to Dollie Cole for giving me greater insight into her late husband, Ed Cole. Thanks to drag racing legend "Dyno" Don Nicholson for the memories he gave me when he raced his Chevys at Detroit Dragway in the early 1960s and for allowing me to spend some time with him at his home discussing his days racing Chevys.

This book would not have been possible without the cooperation of Chevrolet's public relations staff and, in particular, Doris Mitri and Laura Toole. Thanks to GM's media staff, John Robertson, Joe Shively, and Kim Schroeder for the hours of research needed for many of the photos in this book. A special thanks to Floyd Joliet of GM's design staff for his help in locating not only photos, but some of the former Chevrolet employees I interviewed. Thanks to the following car owners for allowing me to photograph their cars for this book: Jerry Wagoner, Steve Halluska, Doug Scott, Carl Weaver, Marvin Wait, Tim Stout, Mike Komo, Jerry Bryant, David Pruitt, Dale Armstrong, Bob Billings, Lisa Foland, and John Green. Thanks to my friends at Dobbs Publishing, Tom Shaw and Greg Pernula, for coming to my rescue when I needed a few specific photos. Thanks to Jonathan Mauk, Linda Johnson, and Tim Mielecki at the Daytona Racing Archives for their research and to the late Leslie Lovett of the NHRA Photo Archives for his help with this project. Thanks to the Late Great Chevy Association and the 1965–1966 Full-Size Chevy Club members for their assistance. Thanks to my buddy Joe Veraldi, photo assistant, key grip, researcher, car wrangler, and good friend.

Many thanks to Motorbooks editor Zack Miller for his patience, coaching, and good advice throughout this project. And last, but definitely not least, to my wife and best friend, Robin. She's my copy editor, organizer, coach, business partner, motivator, and lover. Without her special skills, I'd be lost.

—Robert Genat

Introduction

With the dawn of the 1960s, the age of extravagant visual invention came to an end. It wasn't that the cars no longer had style, it's just that the style was changing. Bill Mitchell, head of GM styling through the 1960s, had his own idea of what a car should look like. His ideas were beautifully interpreted through his design staff. The Chevys of the 1960s were much leaner and conveyed a sense of power and speed—even when they were sitting still. Each car had its own look, but each was unmistakably a Chevy. They were all highly identifiable, designed with a strong front and rear view. Chevy designers wanted everyone to know what was coming at them—and what they were following.

The 1960s was an incredible time to live and work in Detroit. Chrysler, Ford, and General Motors were hiring, but everyone wanted to work at GM and, more specifically, at Chevy. Everyone wanted to be part of the excitement Chevrolet was generating. The Chevrolets of the 1960s looked good and were fast. These cars set sales records 30 years ago and today continue to sell well. The list of collectable cars produced by Chevrolet during the 1960s is lengthy.

The engineering staff assembled by Ed Cole in the late-1950s was ready to take on the challenges of a new decade. Cole and Chevrolet's other general managers through the 1960s, Bunkie Knudsen and Pete Estes, encouraged innovation. Cole, Knudsen, and Estes loved cars and, in today's vernacular—were "motorheads." They valued the efforts the designers and engineers were putting forth to create new and exciting products. The designers and engineers knew their work was appreciated and worked hard for these guys. Cole, Knudsen, and Estes set the tone for the most exciting decade in automotive history.

Chapter 1

1960 A New Direction for Chevrolet

Nineteen-sixty was an exciting year in automotive history. Each of the Big Three automakers (General Motors, Ford, and Chrysler) introduced smaller, more fuel-efficient cars, even though fuel prices were low. This surge in domestic economy cars was intended to compete against increased import sales in the late 1950s. Sales of imported cars from England, Germany, and Japan totaled 550,000 units in 1959. This was a very small piece of the automotive pie compared to the 6.5 million cars domestically produced and sold that year—a small piece of the pie, but one the Big Three didn't want to give up.

In 1960, Ford introduced the Falcon, Chrysler the Valiant, and Chevrolet the Corvair. These cars varied greatly in design, performance, ride, handling, comfort, and economy, and Chevrolet's Corvair was the most innovative of the group and eventually became the most controversial.

Everyone at GM knew they had gone too far in 1959, and the full-size 1959 Chevy's flamboyant wing-like rear fender fins were tamed down for 1960. GM's cars had been pulled in every direction to make the biggest artistic statement possible, and a new sense of restraint was needed for the vehicle design. So while the 1960 Chevrolet retained the unique look of the 1959 Chevrolet, it toned down

Ed Cole's dream was to build a simple, inexpensive car as a second car for a family. The 1960 Corvair offered many firsts to American car design: aluminum rear engine, transaxle, and four-wheel independent suspension. It was also the first car with a unitized body ever built by Chevrolet. *GM Media Archives*

the design's more exaggerated elements, such as the large fins.

Enthusiasts awaiting the introduction of the 1960 Corvette didn't have much to celebrate. Even close up it was barely discernible from the 1959 model. Attempts to wring more horsepower from the 283 engine with aluminum heads were unsuccessful. However the 1960 Corvette introduced new applications of aluminum that were more successful than some of Detroit's earlier attempts.

The Corvair

The dream of two cars in every garage was about to be realized. The stage was set by virtue of a large network of highways constructed in the 1950s. New suburban housing tracts were blossoming around every major city. The American dream of owning a home was becoming a reality for many people. Now, living in the suburbs, people could no longer depend on public transportation to get to work, do their shopping, or run their errands. The family needed the mobility offered by a second car. Ed Cole, general manager of Chevrolet Division, bet the future of the Corvair on the suburban family's need for that second car.

When Cole became head of Chevrolet Division in July 1956, plans were moving forward on what would become the 1960 Corvair. A year earlier, Cole had one of his staff engineers, Maurice Olley, sketch designs relating to engine placement and driveline designs for a smaller car. Studies were also done on a two-piece, aluminum, flat six-cylinder engine. Yet the technology of the day couldn't support the complexity of materials and manufacturing needed to build such an engine for a production passenger car.

The Corvair was developed under the code name of "Holden La Salle II." It was Cole's plan to introduce to the Chevrolet line an exciting new small car that would not hurt the sales of the low-priced, full-size Biscayne. He felt the Corvair would attract a great deal of attention to the entire Chevrolet line and therefore be good for business.

Cole wanted a car that would attract new buyers who were already buying or might otherwise buy imports and domestically produced Larks and Ramblers. The car he envisioned would have a low profile, seat six, and be inexpensive to buy and maintain. A rear-engine car solved many of the package problems inherent in designing a small six-passenger car. With a rear engine there is no need for a transmission tunnel. Three adults could comfortably sit on a bench seat with a virtually flat floor. The rear engine design also shifted weight

The 1960 Chevrolet was the last full-size Chevy designed under Harley Earl's reign. Chevrolet designers maximized sheet metal sculpting to obtain its elegant design. This Impala has added dealer-accessory bumper guards and front-fender gun sights.

away from the front wheels, making power steering unnecessary. Simplicity of design and low maintenance costs were the guidelines of Cole's vision.

Design package proposals for the Corvair were reviewed, modified, and approved in 1956. Within a year, detail drawings were finished for a prototype build. Simultaneously, preliminary sketches were completed on the exterior design.

Some felt the Corvair was revolutionary. Actually, it wasn't. In a sense, Chevrolet simply took the basic design of the Volkswagen and improved it greatly. But compared to the cars Chevrolet had been producing, the Corvair seemed revolutionary. It had an air-cooled rear engine, four-wheel independent suspension, and a unitized body—all firsts for Chevrolet.

Although GM's design head Harley Earl had made his mark in the automotive styling world with some of the most beautiful and flamboyant large cars, he was fascinated by small cars. Following a trip to Europe in the 1950s, he talked more seriously about designing a small car for GM. Occasionally Earl assigned someone in the studios to do a small car study. Designers found they were not able to apply the big sweeps and exaggerations as they had on the large cars. Their design had to be more precise and well-proportioned. Yet Earl's philosophy of long and low emerged in the design of the smaller Corvair.

In August 1957, under Earl's watchful eye, Ned Nickles, head of Experimental Design Studio One, began studies for the exterior design of the Corvair. Carl Renner worked with Nickles on the exterior,

The 1960 Chevrolet instrument panel gauge cluster was carried over from 1959. Its design intent was to resemble a jet fighter's cockpit with each gauge set into its own nacelle. Each of those nacelles had a small "bird beak" at the top to reduce windshield glare when the instrument panel lights were on at night.

In 1960, Chevrolet's fins were slightly tamer versions of the radical arc-shaped fins that first appeared in 1959. Nestled below the gull wing fins on this Impala are three circular lights—two red stop lights and one clear back-up light. This design theme of small pairs and trios of taillights was seen on the back of full-size Chevys for the entire decade.

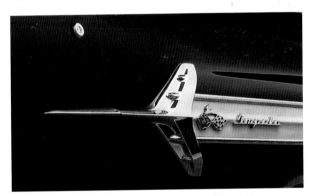

The sides of the 1960 Chevrolets preserved the aviation theme with the placement of a swept-wing jet emblem on the quarter panel. Streaming rearward from the emblem were two chrome vapor trails. The area within the vapor trails was painted a contrasting color to the body.

while Drew Hare worked on the interior. Early clay models incorporated the flying wing rear roof design of the 1960 Chevrolet four-door hardtop. This feature was part of the final production design and gave proportion to the overall design. Another feature of the final design was the upper fender horizontal accent line (sometimes called a rub-strip) that encircled the body. It was one of the tricks designers used to make the Corvair look longer and lower. It was a design feature that would show up later on the 1963 Corvette Sting Ray. The Corvair design never used a false grille in the front. The original design for the front panel was a concave shape with a large horizontal emblem. The Corvair was well-proportioned with striking styling. As with all GM cars, it had to look different but still be distinguishable as a Chevrolet product.

The Corvair was the first car General Motors produced with a unitized body. A unitized body combines the frame and body into a single stressed unit. Fisher Body Division did the engineering under the direction of James Wernig. Fisher Body drew on its European divisions of Opel and Vauxhall for engineering assistance, since they had been building unitized cars since the late 1930s. It was this frameless design that allowed the low profile of the Corvair and created extra package space to seat six adults. The unitized construction also provided a strong, lightweight platform. To prevent premature corrosion of the underbody structural components, Fisher Body coated interior surfaces with a high zinc-based primer prior to assembly. Following assembly, the entire underbody received a coat of paint.

A big drawback to unitized construction is the transmission of noise. Large flat panels, like the floor, were ribbed to add strength and reduce oil-canning, the booming sound made when a panel of sheet metal flexes. A car with a traditional frame structure isolates the passenger compartment from most of the noise and vibration, but the unitized construction, with all major chassis and engine components bolted directly to the body, acts as a conduit for noise. To inhibit noise, Chevrolet added lots of extra body insulation. Rubber cushions were even used in the spring pockets to dampen noise. The motoring public was quite surprised at how quiet the Corvair was inside. Even more surprising was how quiet the Corvair was from the

outside. Everyone expected the "coffee grinder" sound of the Volkswagen when the Corvair drove by, but it had a sound all its own.

The Corvair was developed around its engine more than any other major component, and the Corvair's flat six was a revolutionary design for an American automobile. Undoubtedly, Ed Cole was inspired by the flat six rear-engined M-42 tanks he built while managing GM's Cleveland tank plant during the Korean War. Cole and Harry Barr, who would succeed Cole as Chevrolet's chief engineer, spent their off hours sketching cars—some with rear engines.

Robert Bezinger was the Chevrolet engineer in charge of the Corvair's engine design. Flat sixes had been used in aircraft, but were very expensive to manufacture due to the large amount of machining required. The Volkswagen engine was a flat four. It was determined that a flat four would not develop enough power for the Corvair and the flat six was inherently a smoother running engine.

The design differences between a flat four and a flat six are numerous. The exhaust on the center cylinder of each bank on the six is the real problem. On a four, the exhaust can be ported out of the ends of the cylinders, but the center cylinder exhaust on the six can cause an overheating problem if improperly designed. Chevrolet directed the exhaust down through short exhaust stacks into a manifold. It was then routed forward and turned 180 degrees into a muffler exiting at the rear.

The 1960 Chevrolet was offered in 16 different models with the Impala at the top of the list. The manufacturer's suggested list price on the 1960 Impala sport coupe featured in this ad was $2,468, or $10 less than the same model in 1959.

The exterior of the 1960 Corvette was identical to the 1959 model. Minor changes were made to the interior and chassis, but under the hood is where Chevrolet spent its money. The fuel-injected engine's horsepower was boosted to 315 with the addition of new, free-breathing cylinder heads.

The Corvair's aluminum engine block was rectangular. Its two halves were bolted together at a vertically parting line. Each half of the block had three openings for the individual cylinders. Twelve studs, four for each cylinder, retained the head and its cylinders on each side of the block. No separate main bearing caps were required because the four main bearings were supported entirely by the two halves of the block. Because of limited structural strength with an aluminum block, a forged crankshaft was used. It had six throws arranged in pairs with no counterweights.

Cole wanted aluminum cylinders like those used on Porsches. Porsche heavily chrome-plated the bores of its cylinders so they could withstand the wear of the piston rings. This solution was far

too expensive for the Corvair's intended low price and high volume. Another solution considered was the use of aluminum-silicon alloy, but the technology of the mid-1950s was not in place to support the special machining needed. The Corvair's engine was finally fitted with individually finned cast-iron cylinders similar to the Volkswagen design.

The three cylinders on each side of the engine were sandwiched between the block and cylinder head. The Corvair's heads were made of aluminum and were interchangeable from side to side. The combustion chambers were wedge-shaped with valves actuated by stamped rocker arms. The 1.34-inch diameter intake valves had seat inserts of cast-nickel-steel, and the 1.24-inch diameter exhaust

A bank of bright lights surrounds a line of new Corvairs as they roll off Chevrolet's Willow Run (Yipsilanti, Michigan) assembly plant. Inspectors closely scrutinize each car for missing parts or damage. The Willow Run facility was refitted in 1959 exclusively for the unique assembly requirements of the new Corvair. *GM Media Archives*

valves had cast-chromium-steel inserts. In keeping with the simplicity of design, the spark plugs were installed from the top.

Mounted on each head was a single-barrel down-draft Rochester carburetor. A single carburetor meant long runners to each cylinder bank. These long runners made even fuel distribution almost impossible. The two carburetors were joined by a vacuum balance tube. This tube provided a steady source of vacuum for the choke piston and the Powerglide transmission's vacuum modulator. The carburetor linkage was a nightmarish series of rods, links, and pivots. The *Chevrolet Service News* bulletin of October 1959 declared this system a safety factor. "If for any reason the throttle should be stuck or anything happen to the system, it is possible to reach down and pull the pedal back, closing the throttles." This was a strange warning to be found in a service bulletin. Such a warning should have been emblazoned on the instrument panel.

The Corvair's engine was cooled by an 11-inch diameter fan mounted horizontally on top of the engine. This fan drew in fresh air from the louvers on the deck lid. The fan had a thermostatically controlled restrictor that reduced the flow of air when the engine was cold. At higher engine temperatures it opened up, allowing more airflow. The fan was spun by a single 56-inch-long V-belt driven at a 1.6:1 ratio off a pulley at the rear of the crankshaft. The belt took several bends and twists to drive the fan and the generator. It was kept taut by an idler pulley opposite the generator. The roundabout belt design was not a new invention but an old application from farm machinery. Chevrolet engineers were concerned with durability and worked with fan belt suppliers on groove depth, pulley location, and belt construction. Loss of the fan belt was as detrimental to the Corvair's air-cooled engine as it was to any standard water-cooled engine. In keeping with the simplicity of the overall Corvair design, the fan belt could be replaced without removing any major components.

Unfortunately, the Corvair's engine package weighed in at 366 pounds, a hefty 78 pounds over the target weight. A good portion of that extra weight was due to the cast-iron cylinders. Extra weight means extra cost. Compromises had to be made in the rest of the vehicle to keep production costs down. Some believed these compromises were what hurt the Corvair's performance and market longevity.

A close look at the two blue Chevys floating above the two green ones will reveal some similarities in design. Most obvious is the flight deck roofline and slanted C-pillar shared by both the Impala and Corvair. Both cars share circular taillights and thin horizontal rear bumpers. The front bumpers are also of the thin blade design. One big difference between the two cars is the windshield. Chevrolet's is a full wraparound design; the Corvair's is not. Nineteen-sixty was the last year for the wraparound windshield on a Chevy. *GM Media Archives*

In the spring of 1958, prototype engines were fitted into three test mules, two Porches and one Vauxhall. It was important to prevent the other manufacturers from learning anything about the Corvair tests, so Chevy engineers did their best to conceal what they were doing. The Vauxhall had a large fake front grille to conceal the rear engine design from inquisitive eyes as engineers drove the prototype around Detroit. The Vauxhall even had "Holden Special" nameplates to further confuse the enemy (at the time the Holden was produced by the Australian division of General Motors). Prototype parts were drawn on paper with Holden title blocks, and to make the masquerade complete, those parts were ordered on requisition forms that bore the Holden name.

The 1960 Corvairs were initially offered with a choice between two transmissions, a three-speed manual and a Powerglide automatic. The powertrain layout had the differential sandwiched between the engine and transmission. The clutch on manual transmission cars, or the torque converter on Powerglide cars, was attached to the front of the engine. A shaft drove forward through the differential into the transmission. The output was directed 180 degrees back into the differential.

The Corvair's coil spring front suspension was assembled as a unit with the front cross-member and then attached to the unitized body. The stamped upper control arms were similar to, but smaller than those used on the full-size Chevrolets. The lower arms were a two-piece design. A single channel section stamping attached to the cross-member supported the vehicle's weight. A bar connecting the lower arm and body structure provided rigidity. The strut attachment at the body was insulated with a large rubber bushing and provided the toe adjustment.

The independent rear suspension was a swingarm design. The Corvair's gearbox was rigidly mounted to the chassis. Each equal-length axle shaft was attached to the gearbox by a universal joint. The swingarm design gets its name from the fact that the rear wheels swing through an arc equal to the length of the shaft. The suspension was designed to increase toe as the arm swings in either direction. This slight increase in toe added a desirable understeer effect. Adjustment to the toe was done by shimming the transmission mounts, which then moved the engine/transmission/differential assembly.

The sporty two-door Corvair coupe was a midyear release in 1960. Late in 1960, bucket seats were installed and the Monza was born. This 1960 Corvair coupe was entered in the 1960 Mobilgas Economy Run, a driving event covering 2,061.4 miles between Los Angeles and Minneapolis. A Corvair placed third in the compact class with a miserly 27.03 mile-per-gallon average. *GM Media Archives*

Corvairs were all equipped with 13-inch diameter steel wheel rims that were very wide, 5.5 inches, and mounted on those wheels were 6.50x13 tubeless tires. Tire air pressure was a critical component of the Corvair's handling. Chevrolet recommended 15 pounds per square inch (psi) for the front tires and 26 psi for the rear tires. Steering response could be improved by increasing front tire pressure to between 16 and 18 psi. With improved steering response came increased oversteer. Many Corvair owners failed to maintain proper tire pressure.

Because the Corvair was unlike any car General Motors had ever produced, it had special assembly needs and could not slip into a production line with an existing model. The GM truck plant at Willow Run, Michigan, was expanded for the initial Corvair production. A Fisher Body plant adjacent to the assembly facility built the unitized bodies, and the engines were assembled in Chevrolet's Tonawanda engine facility. During its production run, Corvairs were assembled in several plants across the United States and one in Canada.

Upon its introduction on October 2, 1959, the Corvair was available in a four-door standard 500 model and a deluxe 700 model. The Corvair's list price was $246 less than a full-size Biscayne sedan. Since the Corvair handled so easily without power steering or brakes, those options were not even offered. The automatic transmission was a $135 option, $50 less than the automatic transmission option on full-size cars. A Corvair priced out $500 below the price of a comparably equipped Biscayne and $2,000 less than a fully equipped Impala.

Up until this time, with the exception of Tucker and Cord, there hadn't been much innovation in the automotive world since the 1930s, and the list of Corvair's American automotive firsts is noteworthy: an air-cooled, rear-mounted aluminum engine, a transaxle, and independent rear suspension.

Those who road tested the Corvair loved the light steering and even said it felt power-assisted. They were pleased with the proportions and interior roominess. *Motor Trend* magazine named the Corvair its "Car of the Year" in April 1960. The Corvair was its unanimous choice for "engineering progress with its air-cooled engine, transaxle, and four-wheel independent suspension." But there was one ominous word that appeared in nearly every road test—*oversteer*. This is the tendency for a vehicle's rear end to abruptly swing to the outside when turning. This flaw in the suspension design would come back to haunt Chevrolet.

Very late in the 1960 model year, the Corvair Monza was introduced. It was a sporty upgrade to the two-door coupe. Originally the Monza coupe was a show car destined for the Turin, Italy, car show. It had leather bucket seats, chrome wire wheels, and was a shade of smoky blue-gray inside and out. Ed Cole abruptly delayed the Monza's shipment to Italy and had it brought back into the studio. He wanted to release the Monza as a production car, but first the engineers had to figure out how to reduce its costs.

The Monza's bucket seats, which had been made by sawing the middle out of a standard Corvair bench seat, were retained, but their leather trim was changed to vinyl. The chrome spoke wheels were replaced with sportier wheel covers. Cole's engineers reduced the cost by $600 and created a car with a sporty new image. The Monza was released too late in 1960 to have any impact on the market, but 1961 was another story.

On May 17, 1960, an article appeared in Monterey County, California, newspapers regarding an auto accident and subsequent death of 16-year-old Don Wells Lyford. The teenager was driving his stepfather's Corvair along a twisty, two-lane road and lost control. Don crossed the centerline and collided with a Plymouth in the opposing lane. This accident began the debate over the Corvair's safety.

Chevrolet

The 1960 Chevrolet introduced subtle design changes to temper the dramatic lines first introduced with the 1959 model. Chevrolet stylist Clare MacKichan was responsible for the rear of the 1960 Chevrolet. For 1960, the swooping arc of the rear fins was removed in favor of a more subdued, geometric gull-wing design. Gone, too, were the teardrop taillights, as the 1960 model featured the return of the small, round taillights first seen in 1958. This particular taillight arrangement was carried over through the 1965 models.

The front of the 1960 Chevrolet was styled by Bob Cadaret, who had been impressed with the oval-shaped front end on a Chrysler show car, and he applied the same basic oval shape to the front of the 1960 Chevy. The initial design had headlamps in the center of the grille along with the standard outboard lights—all within the oval grille opening. That design concept didn't last long, though, and the center lights were removed. Cadaret added individual block letters spelling out CHEVROLET across the front of the hood, but in the design review with Ed Cole, Cadaret was told the block letters had to go. The cost of individual diecast chromed letters, along with the extra holes in the hood and assembly, was excessive for a Chevrolet. Cadaret lobbied hard for his design, however, and Cole relented since he liked the overall front end design.

Innovations in the aerospace industry and the automotive designers' fascination with jet aviation were reflected on the exterior design, particularly the side treatment, of the 1960 Chevrolet. Designers Cadaret and MacKichan teamed up to create the suggestion of a swept-wing jet on the side of the quarter panel, and extending rearward on the Impala were two chrome "vapor trails" flowing to the taillights.

The aviation theme was incorporated into Chevrolet's 1960 advertising campaigns. One full-page ad featured an Impala convertible with a sailplane clearly visible in the background. The copy read, "I built my Chevy as a four-wheeled glider with a sailplane's soaring smoothness." In describing four optional accessories, the advertising message goes on to say: "Seatbelts, naturally, and who ever heard of a pilot without his compass? That push-button radio comes in loud and clear at the airfield. And Chevy's throttle holder works like an auto pilot." Corny by today's standards, but it strongly identified owning an Impala with the freedom of flying.

The 1960 full-size Chevrolet rode on an X-frame with a front suspension of short- and long-arm design. Steering was recirculating ball with power assist optional. The power steering pump was moved from the rear of the generator and was belt driven as an independent accessory unit. The rear axle was held up by coil springs, and two lower control arms and one upper arm held the rear axle in place.

Three basic engines were offered in the full-size 1960 Chevrolet. The ancient 135-horsepower six-cylinder engine anchored the powertrain line-up. This mid-1930s design was backed up with a column-shifted three-speed standard transmission, while the three-speed overdrive and Powerglide transmissions were optional.

Ed Cole's magnificent small-block supplied the basic V-8 power. The 283-cubic-inch engine was rated at 170 horsepower with a two-barrel carburetor. It was called the *Turbo-Fire* V-8 and came with a three-speed manual transmission. With the addition of a four-barrel carburetor and *Super Turbo-Fire* decals, the 230-horsepower engine was born. Fuel-injected (FI) 283 engines offered in 1959 for the full-size Chevys were dropped in 1960. The fuel-injection units were difficult for the average owner to keep in tune. The cost per unit of horsepower on the fuel-injected engines was high compared to the 348-cubic-inch engine.

At the upper end of the V-8 scale was the 348-cubic-inch engine. The 348 W block was introduced in 1958 because it was thought the basic small-block had reached its limit in bore and stroke potential. The 348 was available in 5-horsepower-rating increments from 250 to 315. Depending on which 348 was ordered, the customer had the

Ed Cole

Ed Cole

Edward N. Cole was born in Marne, Michigan, on September 17, 1909. His ambition as a young man was to be a lawyer, but his attention shifted to engineering after he worked for a summer for an automotive supplier. In 1930, under the sponsorship of Cadillac, Cole enrolled at the General Motors Institute, and because of his extraordinary engineering talents, he was taken into Cadillac prior to graduation to work on a special project. Cole was promoted within Cadillac through several positions, and when auto production turned to war production in the 1940s, Cole became the chief design engineer for GM's light tank production. At the end of the war he was promoted to chief engineer at Cadillac, and within five years he was back in war production as manager of the Cleveland, Ohio, tank plant, which produced tanks for the Korean conflict.

In the spring of 1952, Cole became Chevrolet's chief engineer. He was brought in to breathe life into what was becoming a stodgy car line that was losing sales. Within a short time, Cole expanded Chevrolet's engineering staff from 851 to 2,900 people. The 1953 Corvette, a project rejected by all other GM divisions, was the slick combination of his engineering of off-the-shelf components and Harley Earl's design. Cole is the undisputed father of the small-block Chevy V-8, a timeless piece of engineering that was designed in 15 weeks. Cole was a brilliant engineer and knew exactly what Chevrolet needed. In July 1956, he was promoted to general manager of Chevrolet Division, a position he held until 1961.

Ed Cole was a tinkerer. It wasn't unusual for the family dining room table to become the workbench for a disassembled carburetor. There wasn't anything Cole wouldn't take apart. His widow, Dollie, said, "The greatest gift I could give Ed was something broken." He spent most weekends with his family at their northern Michigan ranch. While on the road to and from Detroit, it wasn't unusual for Cole to stop and help a motorist who was having car trouble. He looked like John Q. Public in his khaki pants and Chevy station wagon. Never letting on he was a vice president of General Motors, Cole often informed the stranded motorist that had the motorist been driving a GM car, it might not have broken down.

Cole never understood bean counters. He felt they were out of touch with the car-buying public. He traveled extensively to the plants and dealerships to keep his finger on the pulse of the industry. He was well-liked by everyone but the competition—they just respected him.

Ed Cole is the architect of the modern Chevrolet. He invented the small-block engine and assembled a team of engineers at Chevrolet that built some of the greatest cars in automotive history. Chevrolet wouldn't have much history without Ed Cole.

choice of three- or four-speed manual, Powerglide, or Turboglide transmission. The 348 was a unique design often maligned as being a truck engine, but its true potential was yet to be discovered.

Out on the West Coast, a young man by the name of Don Nicholson worked at a Chevrolet dealership running a chassis dyno. Don loved Chevys and loved drag racing. His Chevy had everyone covered at the local drags. Don drove a 1960 Biscayne Utility Sedan with a 315-horsepower 348-cubic-inch engine. With 11.25:1 compression, a solid lifter cam, and tri-power carburetion, it was the biggest engine Chevy offered in a full-size passenger car. He raced for the sheer fun of it, never thinking that one day he would be paid just to show up at a drag strip—driving a Chevy.

Corvette

From the outside, the 1960 Corvette looked like the 1959 model, but underneath its exterior were several small changes that resulted in a marked improvement in the car. Corvette engineers added several aluminum components to reduce weight and improve performance of the 1960 model, and for the first time, more than 10,000 units of a new Corvette were sold. More than half of those sold were equipped with the optional four-speed transmission.

The stiffly sprung, heavy-duty suspension option was deleted from the 1960 RPO list. A new rear sway bar was added, the diameter of the front bar was increased, and the rear suspension's rebound travel was extended by 1 inch. These changes improved the overall handling of the vehicle and produced a smoother ride.

A look under the hood revealed a liberal use of aluminum. The most exciting application was the addition of new fuel-injection heads that were 52 pounds lighter than their cast-iron counterparts. Their new larger intake valves pushed the horsepower of the top-rated FI engine to 315. Unfortunately, the casting technology in 1960 couldn't support the application. Heads with internal flaws made their way into production despite close inspection. A coolant loss and subsequent overheating resulted in irreparable damage to the heads and possible partial melting of the aluminum. Aluminum cylinder head failures prompted Chevrolet to withdraw them early in the production run, replacing them with cast-iron heads with the same dimensions. While the failure of these first aluminum heads was a disappointment to Chevrolet engineers, they continued to work on their design, knowing there was a place for aluminum heads in the future of Chevrolet performance engines.

The fuel-injection units on the 1960 Corvettes were redesigned with a larger plenum, maintaining the height of 12 inches. This new design matched the improved head and intake valve configuration to provide the additional horsepower. The fuelie cars were treated to a new cross-flow aluminum radiator, designed by GM's Harrison Division. Behind that radiator was an optional temperature-sensitive viscous fan drive.

In 1960, U.S. auto manufacturers sold more than 7.8 million cars and trucks, an increase of 17 percent from 1959. Chevrolet's market share in 1960 was 25.8 percent—it was a good year.

This little Corvair is dwarfed under the tail of a Pan Am 707. When the Corvair was initially released in the fall of 1959, it was only offered in a four-door body style. For its small size, the Corvair was well-proportioned and quite roomy inside.
GM Media Archives

The 1960 introduction of the Corvair, Ford's Falcon, and Chrysler's Valiant signaled a shift in the way auto companies would do business in the future. Gone were the flamboyant and expensive styling changes each year. Detroit would instead turn to engineering and simplicity to attract buyers.

A changing of the guard had taken place at General Motors in the late 1950s. Bill Mitchell took over styling at General Motors and Ed Cole became the head of Chevrolet Division. This new guard would manifest itself into an automotive dynasty in the decade ahead.

Chapter 2

1961 The Super Sport Era Begins

Early in 1961 the country was recovering from a mild recession, and sales of durable goods, including automobiles, were down. Toward the end of the year, however, the trend reversed quickly and the entire economy blossomed. Nevertheless, total sales of General Motors cars and trucks were 14 percent below 1960 levels, and overall auto industry sales were down 15 percent. The exciting news for GM in 1961 was that fourth-quarter sales figures were at near-record levels.

In 1961, General Motors recognized the auto industry's potential as an emerging global enterprise. Prior to World War II, some 1.7 million vehicles were sold per year in nations outside North America; in 1961 that figure had grown to 6.7 million. Yet that overseas demand for vehicles was being met by manufacturers within each country.

Following World War II, GM anticipated this trend and bought into Vauxhall in England and Opel in Germany. In 1961, 14 percent of General Motors' earnings were generated by these and other overseas investments. Two-thirds of each dollar earned overseas benefited GM's domestic coffers,

The rear of the 1961 Corvette was redesigned with the essence of Bill Mitchell's Sting Ray racer. The rounded deck and rear fenders of the 1960 Corvette gave way to this bobtail design. Seen for the first time were four circular taillights, a Corvette styling cue that would be carried throughout the decade. When ordering a new Corvette in 1961, the hardtop typically came in the car's body color. This attractive Jewel Blue Corvette is unique because its Ermine White hardtop matches its coves.

The 1961 Chevy was the first full-size Chevrolet to be designed under the direction of Bill Mitchell. The 1961 Impala and Bel Air sport coupes both had this stylish roofline known today as the "bubbletop."

and the remaining one-third was reinvested into its overseas plants. In the 16 years following World War II, General Motors' income from overseas investments totaled $5 billion. In domestic sales for 1961, General Motors produced a total of 3.1 million cars and trucks, 1.6 million of which were Chevrolet passenger cars.

Nineteen sixty-one was a transitional year for Chevrolet. Ed Cole moved up in the General Motors organization and Bunkie Knudsen came over from Pontiac to be the new general manager of Chevrolet. All three Chevrolet models underwent major changes that affected the direction of the product. The Corvair was given a sporty image with the addition of bucket seats, four-speed transmission, and more horsepower. The Impala's redesigned exterior took on a sporty look with the SS option. Chevrolet's marketing department also realized how much free advertising it could get by winning a few races and loaning a few cars to enthusiast magazines to road test. The Corvette also looked into the future

by taking a piece of the Stingray racer design. Things were starting to cook at Chevrolet.

Chevrolet

In the fall of 1960, Chevrolet introduced its new, slightly smaller, full-size Chevrolet for 1961. The rear wings were only imaginary, outlined by chrome along the quarter panel and up over the deck lid. The 1959 and 1960 Chevrolets were criticized for their low rooflines and minimal headroom. Some critics said Chevrolet, with its lack of headroom, was responsible for the trend away from men wearing the snap brim hats that had been so popular in the 1950s. Headroom and overall vehicle height on the 1961 Chevrolet was increased by 2 inches and the overall length was shortened by 1.8 inches, all of which represented a big change from the days of Harley Earl.

The 1961 full-size Chevrolet body was completely new, and it was the first full-size Chevrolet designed under the direction of Bill Mitchell. Clare

The 409-cubic-inch W engine was a late introduction into the 1961 line-up. Only 142 409-powered SS Impalas were built, with several going to the major auto enthusiast magazines for road tests. The 409 was only available with the SS option, although several racers bought their engines at the Chevy parts counter. The 409 looked identical to the 348-cubic-inch engines, except the valve covers were painted silver. This particular 1961 SS is exceptionally rare because of its optional factory air conditioning.

The 1961 Super Sport interior featured a passenger assist bar on the right-hand side of the padded instrument panel (a required option). Super Sports with a four-speed received a chrome trim plate at the base of the shifter. A Sun 0–7,000 rpm tachometer (with no redline) was attached to the steering column. Clearly visible on the seats are the design pads sewn in under the seat covers to give some contour to Chevy's traditional slab seats.

MacKichan came up with the design for the bob-tailed rear deck. Young Chevrolet designer Paul Deesen worked out the front end and was able to carry over the Chevrolet name in block letters on the hood. It was a very conservative design compared to its predecessor. The raised roofline and resultant redesign of the A-pillar eliminated the wraparound windshield used on 1958 through 1960 Chevrolets.

The interior of the 1961 Chevrolet was all new, and the most dramatic change was the redesigned instrument panel. The jet cockpit design of the 1960 model's instrument panel was set aside for a more upscale look. The speedometer was long, horizontal, and deeply recessed into a large cove. At the base of the steering column were three small circular bezels which housed the gas gauge, temperature gauge, and a clock, if so optioned. The glovebox was moved to the center of the instrument panel, making it easily accessible by both driver and passenger.

Chevrolet seat cushions and backs had been traditionally flat, almost bench-like in design. Due to cost, any additional detailing of the seats was reserved for Pontiac, Buick, Oldsmobile, or Cadillac. One trick used to detail the seats on these upscale cars was to sew in a design pad (strips of foam under the upholstery) to give the seat shape. For the first time in 1961, the seats of the Chevrolet Impala featured a design pad. This pad consisted of four 3-inch wide strips of 1/2-inch-thick foam sewn into the front seat cushion. These strips gave the cushion

shape and definition. This design was worked out by Chevrolet interior designer Blaine Jenkins and was approved by Ed Cole at one of his 7 A.M. visits to the Chevrolet studios. Total cost per Impala for the design pad was $1.50. Cole liked to put the money where it would be seen when the potential customer looked into the car at the dealer's showroom.

The new 1961 Chevrolet body rode on the same X-frame as the 1960 model. The only change under the car was the relocation of the gas tank from under the trunk floor to the area between the trunk and rear axle. This allowed the trunk to have a very deep well, a selling point against Ford's shallow luggage compartments. The relocation of the gas tank required a move of the gas filler from behind the rear license plate (as it was on the 1960 model) to a gas filler door on the left quarter panel.

The Impala SS was added to boost the sporty image of the Impala. The Super Sport option was a late addition to the line-up for 1961—so late it didn't make it into the sales brochure for new car introductions, and a special brochure was printed and shipped to dealers after the fall introduction of new cars. This brochure featured an illustration of a four-door Impala hardtop with the Super Sport trim. Rough estimates show 450 Super Sports were sold in 1961, and there were probably no four-door models.

The Super Sport was a trim option added to an Impala with a high-performance engine and chassis combination. Special SS body trim included deck lid and quarter panel emblems with a red SS overlay. Super Sport wheel covers used the standard 1961 wheel cover with a tri-bar simulated knock-off spinner attached. All Super Sport wheel rims were painted black regardless of body color. It was on the 1961 Super Sport that Chevrolet first used thin-line white sidewall tires (8.00x14).

The Super Sport interior consisted of the Impala trim, plus a few extras. A Corvette-style passenger assist bar was added on the right side of the padded instrument panel, and a chrome Impala SS emblem was set into the left side. Behind the steering wheel rested a Sun tachometer with a range of 0 to 7,000 rpm with no redline indicated. The tach was positioned at the 10 O'clock position on the steering column, allowing full view of the speedometer and temperature gauge. This is arguably the best-looking factory-installed tachometer ever. Because the SS program was initiated so late, the tachometer was an add-on and was not designed into the instrument panel. At the base of the four-speed shift lever was a chrome trim plate decorated with crossed flags.

The problem with the 1961 Super Sport option was the high cost of the mandatory options the customer had to order just to get the SS trim. The buyer had to order one of three high-performance 348-cubic-inch engines or the 409. Transmissions were limited to a four-speed manual or high-performance Powerglide, which was available only with the 305-horsepower 348. Also required were heavy-duty front and rear springs and shocks and thin-line white sidewall tires. Power-assisted steering and brakes had to be added to the list as well, and finally, a padded instrument panel and tachometer rounded out the list of "must haves." These required options boosted the sticker price of the Impala by more than $600 before the $50 SS option could be added. Add a radio, heater, and a few other options and that Impala SS could sticker over $4,000—a lot of money in 1961.

In 1960, Chevrolet sold 10,000 cars with high-performance engines. In 1961, knowing that cubic inches equal cheap horsepower, Chevrolet introduced the 409. It was a bored-and-stroked version of the 348, rated at 360 horsepower at 5,800 rpm. The W block, as it was known, had some distinct design advantages, as it was the only production engine with fully machined combustion chambers. Each cylinder was surrounded by six head bolts for a total of 18 per side, while other V-8 engine designs had only 10 or 12. The W block's deck angle was 74 degrees instead of the traditional 90 degrees, and this unique deck angle created oval-shaped cylinder bores. The area of any cylinder cut on an angle will be greater than that of one cut at a right angle. The combination of the additional area and the staggered valve arrangement provided room for larger valves. Externally, the 1961 409 looked similar to the 348, except the rocker arm covers were painted silver instead of Chevrolet-engine orange.

Since the 348 W block would not accommodate the changes needed for the 409's larger bore and stroke, a new block was cast for the 409. The 409 had larger crankshaft counterweights, requiring changes to the bottom of the block to make room for all those spinning parts. The 3/16-inch larger bore brought the total cylinder diameter to 4 5/16. The 348 block could not accommodate this large bore without breaking into the water passages.

The forged crankshaft, with its larger counterweights, weighs 67 pounds, 8.2 pounds more than the 348 crank. A 409 crank can be distinguished quickly from that of a 348 by the extra counterweight on the flywheel flange. The new rods were 1/8-inch shorter, yet while shrinking in length, they were larger in girth as extra material was added along their entire length. The weight increased over the 348 rod by 2.45 ounces to a total of 27.16 ounces. The 409 pistons were forged aluminum with an advertised compression ratio of 11.25. All 1961 409s were delivered with dual head gaskets, effectively dropping the compression ratio to 10.3. This was done to make the 409 a little more civilized for the street.

The 409 camshaft had an intake duration of 317 degrees and exhaust duration of 301. Both intakes and exhausts work at a lift of 0.440. The valve lash settings were 0.008 for intake and 0.018 for exhaust, both set hot.

The 409 heads were the same as the 348 heads with two exceptions: They had slightly larger pushrod holes to accommodate the larger, 3/8-inch diameter pushrods, and the valve spring bosses were enlarged to accept single springs with dampeners. The intake valve diameter was 2.060, and the exhaust diameter was 1.720. There is no combustion chamber in the head. A small amount of volume is created by the head gasket and the valve seats, but for all practical purposes the head is flat and the combustion chamber is in the block.

The 409's aluminum intake manifold is the same as the 1961 340-horsepower 348 engine, with one exception: The throttle bores had to be opened up on the primary side to accept the larger Carter AFB. The ignition was a dual-point Delco with no vacuum advance. The exhaust manifolds were cast-iron and had a 2 1/2-inch opening to the mufflers.

The 1961 Chevrolet was shorter and narrower than its 1960 counterpart. The only exterior dimension that grew was overall height. This 1.5-inch increase was added to interior headroom.

In the summer of 1961, Chevrolet engineering created a service package for the 409. It contained new, larger-port heads, a longer duration cam, a dual four-barrel aluminum intake manifold for the drag racers, and a single four-barrel intake for NASCAR. This service package increased the output by 50 horsepower and was standard on the first production 409 Chevys in 1962. Drag racers using this package were placed in Optional Super Stock (OSS) class—a class created to cover the flood from all manufacturers of new factory parts not in regular production. In 1962 NHRA designated this class Factory Experimental (FX).

In 1961, the 409 was intended to be available only in the Super Sport models. Some 142 Chevys with the 409 engine were produced in 1961, and most of these production vehicles went directly to racers or to lucky journalists for magazine road tests. Chevrolet received maximum return on its press car investment as magazines gave the new 409 SS rave reviews.

Motor Trend tested a 409-powered 1961 Super Sport in its September issue. It was driven with two different rear-end ratios, a 3.36 posi and a 4.56 posi. The Impala SS (with the 3.36) did the quarter in 15.31 seconds at 94.24 miles per hour, and gas mileage was 10 to 14 miles per gallon. Using the 4.56, the times dropped to 14.02 seconds and the speed increased to 98.14 miles per hour. Not bad for

a full-size production car with a closed exhaust and narrow bias tires. The 1961 409 SS was a resounding success, and it drew attention to Chevrolet's high-performance engine program and to the new Super Sport trim package.

The new 409 engine quickly made a name for itself. West Coast Corvette tuner Bill Thomas regularly used Service Chevrolet's top chassis dynamometer man, Don Nicholson, to tune his fuel-injected roadracing Corvettes. Thomas was aware that Nicholson had been cleaning up on the drag strips with his 1960 Biscayne and was a whiz with the 348 engine. Thomas arranged through Vince Piggins, Chevrolet's high-performance rep, for Nicholson to get one of the first 409 engines for the upcoming NHRA Winternationals in Pomona, California. The engines were not available through regular dealer channels, nor were they available off the showroom floor. The 409 was legal for NHRA because Chevrolet had provided the Automobile Manufacturers Association (AMA) with specs for the engine.

Nicholson purchased a new, white 1961 Impala two-door hardtop with a 348-cubic-inch engine, four-speed transmission, and positraction rear axle. This Impala had no radio or heater. Nicholson had Jerry Jardine build a set of headers in anticipation of the soon-to-arrive 409. The 348 and 409 had the same external dimensions, making Jardine's task

25

Super Sport Impalas were the only Chevys available in 1961 with thin-line white sidewall tires. Bolted to the center of the full wheel cover on the Super Sport was a tri-bar spinner. Small SS emblems were attached just ahead of the Impala script on the quarter panel.

easy. Just days before the race, the new 409 engine was delivered to Nicholson, and it was torn down and carefully reassembled, but not bored to the NHRA-allowable .060 over. On the Wednesday before Friday's opening of the 1961 Winternationals, Nicholson was lowering the 409 into his Impala. Thursday was spent with final installation details, and that evening Nicholson got some much-needed sleep. Before going to bed, he gave the keys to a friend and asked him to drive the car around town all night to break everything in. Nicholson wanted all the gears and wheel bearings seated-in before the race.

Frank Sanders from Phoenix, Arizona, was the only other competitor at Pomona with a 409-powered Chevy, one also set up by Nicholson. The Ford camp was there with its new 401-horsepower tri-power 390-cubic-inch engine. Class eliminations were held on Saturday for the Super Stock class, and the final run came down to Sanders versus Nicholson, with Sanders taking the class win. On Sunday's final run for Stock Eliminator honors, the pairing was the same, but the outcome was reversed with Nicholson taking the victory. Nicholson ran the quarter-mile consistently in the mid- and low-13s at speeds as high as 108 miles per hour. It was an auspicious debut for the new 409 engine.

News of Nicholson's and Sanders' domination

The "wings" on the 1961 Chevrolet were merely body character lines that ran from the quarter panel across the rear of the car, ending in a crisp V on the back of the deck lid. The Impala featured six circular taillights, while the Biscayne and Bel Air had four each. Wide whitewall tires were available as an option. *GM Media Archives*

of Pomona's Winternationals traveled quickly through the pits at Daytona, where preparations were underway for the 1961 Daytona 500. Once on the track at Daytona, the 409s were a disappointment. They were only slightly faster than the previous year's 348s and were no match for the 390-cubic-inch Fords and 389-cubic-inch Pontiacs. Some felt it was the shape of the Impala's rear deck that spoiled the aerodynamics. The story of the race was Pontiac finishing one, two, and three, while the new 409s were never a threat. Two 409s ran well enough to finish a distant seventh and eighth. NASCAR's short

The Impala two-door sedan was offered only in 1961. It carried the same flat-topped roof body as the 1961 Biscayne and Bel Air two-door sedans. The Impala two-door sedan with a V-8 listed for $2,312, $56 less than the Impala Sport Coupe. *GM Media Archives*

tracks were kinder to the new Chevy engines where their short-stroke design could be best utilized for quick acceleration off the turns.

Back on the drag strips, 409s continued their dominance. Spectators were drawn to the tracks for match races between their favorite drivers at the wheel of their favorite cars. Tow money (cash for just bringing your car to a race) was now being paid to 409 Chevy drivers by promoters. New car dealers began providing limited sponsorship money to certain racers to have the name of their dealerships on the side of the cars. Winning on Sunday and selling on Monday was happening across the country at Chevy dealerships. It was the beginning of a trend.

Corvair

Corvair sales in 1960 accounted for only 250,000 units, 50,000 short of Ed Cole's predicted

This ad for the 1961 Super Sport ran in the May 1961 issue of *Hot Rod* magazine. The 409 engine was available at the time this ad ran, and its 360-horsepower rating is listed as one of the engines available. The Impala convertible (inset) is lacking the tri-bar wheel cover spinners.

In September 1961, local United Auto Worker (UAW) strikes disrupted General Motors' automotive production for three weeks. GM employed more than 300,000 hourly employees in 1961, and their yearly compensation totaled nearly $1.9 billion. The average GM hourly worker earned $3.15 per hour, well above all other U.S. manufacturing employees.

In the fall of 1961, a new three-year contract was negotiated with the UAW and other unions representing GM's hourly employees. An annual improvement factor and cost-of-living adjustments provided UAW workers with three years of pay increases.

The new agreement guaranteed for each UAW employee, in each year of the three-year contract, an increase of 6 cents per hour, or 2 1/2 percent of the employee's base wage, whichever was higher. Quarterly cost of living adjustments were tied to the Consumer Price Index. A Supplemental Unemployment Benefit (SUB) plan augmented state unemployment insurance for those UAW workers laid off or asked to work fewer than 40 hours per week. Under the new agreement, SUB payments could last as long as 52 weeks. In 1961, General Motors contributions to the SUB pool equaled $15 million and GM picked up the full cost for hourly workers' health insurance at a cost of $36 million. In December, GM reached a three-year agreement with its Canadian UAW workers that paralleled the U.S. plan. More than 100,000 salaried GM workers received salary and benefit increases that equaled the UAW's gains.

For each dollar of revenue in 1961, General Motors paid 30.5 cents to employee payrolls and benefit plans. That figure was only exceeded by the 47.5 cents paid to suppliers. Shareholders received only 6.25 cents per dollar of revenue.

The general philosophy of the American auto companies was to pay the unions what they wanted. The Big Three had a lock on the American automotive market, and it was financially prudent to capitulate to the union rather than face a strike. With no outside (foreign) competition, all that was needed was an increase in the price of the product to cover the pay increases. It was a golden time for the UAW.

---

sales of 300,000. The Corvair was outsold by the more conservative Falcon, which, with its traditionally laid-out front engine and live rear axle, was more readily accepted by the consumer. The Falcon was also $126 cheaper than the Corvair, a wide gap between vehicles in the same price range. Many potential buyers felt the Corvair was too exotic and could not be repaired easily. Rumors of carburetor icing and fan belt troubles bolstered those beliefs, but Cole and his engineers were not about to let the Corvair die.

The shot in the arm Corvair needed was the Monza, which was introduced too late in the 1960 model year to have much of an impact on the market that year, with only 11,926 Monzas sold. In 1961 the Monza package was extended to include the four-door sedan. Other late upgrades to the Corvair included an optional 95-horsepower engine and a four-speed manual transmission. These new elements transformed the Corvair from an "econobox" into a sports car.

Enthusiast magazines raved about the sportiness of the 1961 Corvair Monza. They loved the bucket seats and the four-speed transmission. The extra horsepower was an added bonus. The Corvair finally found its market.

In 1961, Chevrolet released the Corvair Lakewood station wagon. For a small car, it had an amazing 58 cubic feet of cargo space behind the front seat and 10 additional cubic feet under the hood. Its main rival, the Ford Falcon, had 80 cubic feet of load space and stickered for only $2 more than the Corvair. To get to the Lakewood's engine, you had to remove the rear load floor. The engine location radiated a lot of heat to the interior. The little wagon didn't sell very well, tallying only 25,000 units out of a total of over 300,000 1961 Corvairs.

Following the death in 1960 of young Don Lyford in a Corvair, Don Harney, a former law partner of Lyford's father, started to investigate the safety record of the Corvair. Harney was encouraged by a discussion he had with a Los Angeles police officer who claimed that within six months he had seen six Corvairs flip out of control. In June 1961 the law firm of Harney, Ford, and Schlottman filed suit against General Motors and two Chevrolet dealers. It was Harney's contention that General Motors had been negligent in the design of the Corvair and that dealers were liable for selling a defective product. It would be five years before the case reached Los Angeles Superior Court. During that time, Harney amassed an additional 48 lawsuits alleging the Corvair's defective design.

"Dyno" Don Nicholson won the NHRA Winternationals driving this 1961 Impala powered by the new 409 engine. Nicholson received his engine from Chevrolet prior to its official release to the public. Nicholson spent the summer match-racing across the country. Late in the year, Chevrolet gave him new heads and a dual-quad intake manifold, components that would be standard on the 1962 409s rated at 409 horsepower. To cope with the many new factory parts, NHRA created an Optional Super Stock (OSS) class, the forerunner of the Factory Experimental (FX) class. *NHRA Photographic*

Corvette

The 1961 Corvette stunned its fans. The revised rear styling brought both criticism and praise as Chevrolet made a major change to the Corvette's evolutionary styling trend of the previous five years. The softly curved deck lid and rear fenders were reshaped. The rear deck was lifted and a sharp character line traveled from the top of the rear wheel openings around the back. Set into the angular rear panel were two pairs of round taillights, a Corvette styling feature that remained throughout the decade. Below the taillights were chrome bumperettes, with a license plate bezel recessed in between, and the exhaust exited below the rear panel. The new rear design was taken from Bill Mitchell's Stingray racer and the XP-700 show car.

The Corvette's side coves were still there and, for an extra $15, could be painted Sateen Silver or Ermine White, depending on the body color. This was the last year for the color cove option. Around the front, the grille was now a horizontal mesh insert replacing the 1960 Corvette's large teeth.

The 1961 Corvette's mechanical changes were few. The top engine was a fuel-injected 283 rated at 315 horsepower, but this was the last year the 283 was offered in the Corvette. The aluminum radiator was standard on all engines and the four-speed transmission was now housed in an aluminum case. The transmission tunnel was shrunk, giving the driver and passenger a bit more legroom. The all-vinyl interior was offered in red, fawn, blue, and black.

Hot Rod magazine tested a 1961 fuel-injected Corvette for its July 1961 issue. Writer Ray Brock had a tough time getting his 6-foot, 2-inch frame in and out of the car with the hardtop in place. He also felt a bit cramped while driving, but admitted he was taller than average. His test car was hard to start when hot, as were all fuel-injected Corvettes. Overall, Brock liked the car, especially the close-ratio four-speed transmission. "A short session running up and down through the gears of a Corvette four-speed," he wrote, "gives anybody a new outlook on life."

Chapter 3

1962 The Horsepower Race Heats Up

Nineteen sixty-two was a good year for the economy and for the American automobile industry, whose sales totaled 6.7 million units. Sales of General Motors cars and trucks were 34 percent above 1961 levels and 4 percent above the record year of 1955. Growth in the small car arena was strong, representing 28 percent of all GM cars sold. Earnings per share of common stock rose to a record $5.10!

In 1961, the AMA recommended that by 1963, all cars and trucks manufactured in the United States should be equipped with a crankcase blow-by device to reduce airborne pollutants. In 1961, General Motors added a new smog chamber in its Tech Center to simulate urban air quality. GM consulted with the Sloan-Kettering Institute for Cancer Research to analyze the harmful pollutants emitted by the internal combustion engine. From this research, GM's AC Spark Plug Division developed the positive crankcase ventilator valve, better known as the PCV. It was the first device to be approved by the California State Motor Vehicle Pollution Control Board. It was a simple valve that funneled crankcase fumes into the intake manifold to be burned in the combustion chamber. The PCV was required on all 1962 cars registered in California

Chevy Impalas in 1962 featured a single thin body molding with a color-keyed insert. The color of the insert was coordinated with the color of the interior trim. The crossed flags above the V on the front fender indicate a 327-cubic-inch engine under the hood. Thin-line white sidewall tires, available only on the Super Sport in 1961, were the only optional whitewall tires in 1962.

Two Chevrolet Sport Coupes. On the near side is a 1961 Impala and on the far side is a 1962 Impala. The difference in rooflines between the two is very apparent. The 1962 Impala roof is more upright with transverse creases toward the rear, giving the roof the look of a convertible top. The sporty bubbletop roof on the 1961 model was used only on the 1962 Bel Air sport coupe. As different as these two bodies appear to be, the doors are interchangeable.

and optional in all other states. The PCV added an additional $5 to the cost of a new Chevy.

In 1962, seatbelts for the front seat driver and passenger were optional on all Chevrolet passenger cars except the Corvette, where seatbelts had long been standard equipment. To retain the belts, reinforcements with threaded inserts were welded into the floor pan. The delivering dealer need only slit the carpet and insert eye bolts to attach the belts. Chevrolet customers had the option of buying belts for both the driver and passenger for $17.50, or for the driver only for $9.50.

It was reported that 4.5 million cars in the United States were equipped with seatbelts in 1962, approximately 7 percent of all cars on the road. The popularity of seatbelts brought about a rash of substandard aftermarket seatbelts. These belts had inferior webbing and buckles and were often improperly installed. In Congress, the House Safety Subcommittee drafted a bill regulating all seatbelts sold for interstate commerce.

Virtually every Chevy performance lover in 1962 wanted one of these dual-quad 409 engines in his or her car. The 409 engine with dual-quads added an additional $376.65 to the sticker price.

Chevrolet offered two versions of the two-door sedan in 1962, the Bel Air shown here and the Biscayne. The Bel Air offered a higher level of exterior trim with the long, horizontal, anodized aluminum side molding and thin aluminum molding that outlined the rear of the deck lid and taillights. The Bel Air's interior featured full carpeting and rear seat passenger armrests. The small hubcaps were standard on both models, with full wheel covers optional. The crossed-flag emblem positioned over the small numbers *409* on the front fender indicate this sedan is powered by one of the two 409-cubic inch engines Chevrolet offered in 1962.

Available as an option for the first time in 1962 was the Delcotron, more commonly known as an alternator. General Motors shied away from using the term alternator when describing its AC generator, because Chrysler was already using that term. Delcotrons were available in three output levels: 42-amp, 52-amp, and 62-amp. It was an expensive option, costing as much as $90 for the 62-amp version.

In 1962, the infamous Turboglide transmission was no longer on the option list, a passing for which few tears were shed. The transmission news of note was the new lightweight aluminum case for the Powerglide. It was available on all but the six-cylinder and the 283-powered full-size passenger cars. Those two engines used the older cast-iron version.

Chevrolet produced an odd engineering mix of vehicles for 1962. The full-size Chevrolet was a conventional large car with steel body and X-frame. The Corvette was a fiberglass body with a perimeter frame. The Corvair was a unitized steel body with a rear engine. The new Chevy II was a unitized steel body with a conventional driveline. Something for everyone. But the engineering costs were high and an effort was made to use certain parts on as many cars as possible.

Chevy II

In June 1961, *Motor Trend* magazine reported that Chevrolet was working on a new class of automobile to fit in between the full-size Impala and the compact Corvair. The new car was said to ride on a 115-inch wheelbase, to be powered by a four-cylinder engine, and to have single-leaf springs supporting the car's rear axle. *Motor Trend* also reported that the code name of this new Chevrolet was H-35, often shortened to just H. In the fall of 1961, Chevrolet introduced the H car to the public as the Chevy II.

Ed Cole gave the green light for the Chevy II project in the summer of 1960. Harry F. Barr, who directed the engineering for the Chevy II, had worked with Ed Cole on the small-block V-8 as Chevrolet's assistant chief engineer in the early 1950s. On July 2, 1956, Barr was promoted to the position of Chevrolet's chief engineer.

The 1962 Chevy II was designed and built to compete against Ford's Falcon. It had a conventional driveline like the Falcon, but was slightly larger overall. Chevrolet touted the Chevy II as being "Thrifty, Nifty and New." Chevy IIs were powered by either a four- or a six-cylinder engine. This GM photo shows a small V emblem on the front fender, denoting a 283-cubic-inch V-8. This engine was not offered in the Chevy II in 1962, although dealer kits were available for a V-8 conversion. *GM Media Archives*

A goal for the new Chevy II was to provide transportation for the American family at a reasonable cost. It was to be inexpensive to purchase, economical to operate, and easy to maintain. The Chevy II was to have maximum interior package size without a bulky exterior. The Corvair lacked interior roominess and had little luggage space. The Corvair was also being outsold by the more conventional Ford Falcon. The new Chevy II compared to the Falcon in roominess, body construction, engine size, and driveline layout.

The new Chevy II's body was a remarkable piece of engineering, and no expense was spared in its design and construction. It consisted of two unitized assemblies bolted together at the cowl. The design of the passenger compartment section was well thought out, featuring box sections in the roof rails and a double cowl. Underbody components were zinc-coated prior to priming and sealing. The front-end structure was a welded assembly of front rails, fender aprons, radiator support, and crossmembers. The front-end assembly was attached to the passenger compartment in four widely spaced locations with 14 bolts. It was a design that had been used years before on the Cord. This two-piece body construction was easy to manufacture and service. The benefits of this design outweighed the small increase in total vehicle weight.

The Chevy II suspension was a conventional design, with two exceptions. The front springs were mounted on top of the upper control arm, and the rear suspension was supported by single leaf springs. The placement of the front springs allowed the load to be more evenly distributed throughout the unitized front structure. Only the station wagon models were equipped with a front sway bar. One unique feature of the front suspension was the way in which alignment was done. Threaded bushings with eccentric washers could be turned for caster adjustment and a threaded strut rod similar to the one used on the Corvair was used for toe-in. This simple design eliminated the need for shim packs.

The rear suspension was a conventional live axle design supported by two single leaf springs, which

The wide range of models Chevrolet dealers had to offer in 1962 is displayed in this three-car photo. Pictured to compare the overall length are (top to bottom) the 180-inch Corvair four-door sedan, the 183-inch Chevy II four-door sedan, and the 209.6-inch Impala Sport Sedan. These cars were representative of the 32 passenger-car models Chevrolet offered in 1962. *GM Media Archives*

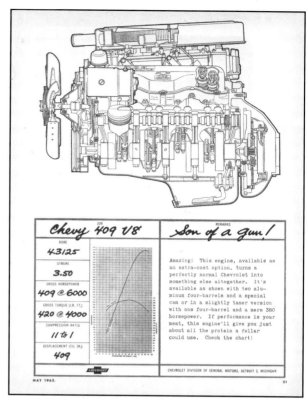

Chevy 409 V8		Son of a gun!
BORE **4.3125**		REMARKS Amazing! This engine, available as an extra-cost option, turns a perfectly normal Chevrolet into something else altogether. It's available as shown with two aluminum four-barrels and a special cam or in a slightly tamer version with one four-barrel and a mere 380 horsepower. If performance is your meat, this engine'll give you just about all the protein a feller could use. Check the chart!
STROKE **3.50**		
GROSS HORSEPOWER **409 @ 6000**		
GROSS TORQUE (LB. FT.) **420 @ 4000**		
COMPRESSION RATIO **11 to 1**		
DISPLACEMENT (CU. IN.) **409**		
MAY 1962.		CHEVROLET DIVISION OF GENERAL MOTORS, DETROIT 2, MICHIGAN 21

Chevrolet placed this full-page advertisement in the May 1962 issue of *Hot Rod* magazine touting the power of the 409 engine. General Motors agreed in theory with the American Automobile Manufacturer's (AMA) ban on racing, but aggressively continued its program to develop high-performance engines. In general, the bulk of Chevrolet's advertising was not performance-oriented, whereas Ford and Chrysler aggressively promoted their high-performance vehicles.

Chevrolet advertised as Mono-Plate springs. Criticized by some as being unsafe, the single-leaf design was effective and proved to be as safe as multileaf rear springs. Both front and rear spring rates were relatively soft to sustain a boulevard ride. Thirteen-inch diameter wheels were standard with either 6.00 or 6.50 size tires, depending on the model.

Two engines were available for the Chevy II and both were brand new. Chevrolet engineers designed a new 153-cubic-inch, 90-horsepower inline four-cylinder engine and a new 194-cubic-inch, 120-horsepower inline six-cylinder engine. These engines shared many internal components and both blocks were machined on the same newly installed, automated line. Both blocks featured a bulkhead on each side of every connecting rod. This design produced a sturdy block with five main bearings for the four-cylinder and seven main bearings for the six-cylinder. The cylinder heads for both engines resembled those on the small-block V-8. The heads used a wedge-shaped combustion chamber, stamped rocker arms with pressed-in studs, and hollow pushrods to carry oil to the valvetrain. A single-barrel carburetor was used for both engines, with the six-cylinder's carburetor having an automatic choke.

Only two transmissions were offered in the Chevy II, a column-shifted three-speed manual and a Powerglide. The manual transmission was essentially the same as that offered in the full-size cars. The air-cooled

aluminum Powerglide was similar to the Corvette's Powerglide. The outer surface of the torque converter housing had air scoops to increase cooling airflow.

The 1962 Chevy II was offered in a full range of body styles. The 100 and 300 series offered a two-door sedan, a four-door sedan, and a four-door station wagon. The deluxe 400 Nova series offered a two-door sport coupe, a two-door convertible, and a four-door station wagon. The 400 series Novas were not available with the four-cylinder engine.

The list of options for the Chevy II was extensive for an economy car. Power steering and brakes were offered, but neither was needed. Factory air conditioning, along with a long list of heavy-duty chassis components, was available, and for an extra $50 bucket seats could be installed in the sport coupe or convertible. The convertible did not have a power top, but the mechanism was well-engineered, as it was spring-loaded, enabling one person to raise or lower the top.

At the 1962 NHRA Winternationals, Don Nicholson backed up his 1961 Stock Eliminator win, this time driving a 409-powered 1962 Bel Air Sport Coupe. This body style was the favorite of both NASCAR and NHRA racers, because of its lighter weight and slicker roofline. Crafty drag racing promoters across the country sponsored weekly match-race events pitting 409 Chevys, like Nicholson's, against Fords and Mopars. *NHRA Photographic*

Car Life magazine thoroughly tested six different Chevy II models and raved about them all. The magazine's only recommendation for sheer driving enjoyment was to order the optional six-cylinder engine. *Car Life* liked the car so well, the magazine awarded the Chevy II its 1962 Engineering Excellence Award. "We think it represents a return to sensibility in terms of basic transportation," the editors said about the Chevy II. "It is a car of reasonable size, adequate performance and simple elegance."

The paint was barely dry on the first production Chevy IIs when Chevrolet Engineering made a complete kit available to install the small-block V-8. The March 1962 issues of both *Hot Rod* and *Motor Trend* magazines detailed the installation of a small-block using over-the-counter Chevy parts. Chevrolet had the vision to initially engineer the car with enough room to accept the V-8 without any major modifications. The balance of the parts needed were right there in the kit—new radiator, metallic brakes, a floor extension for the four-speed shifter, and molded radiator hoses. Don Nicholson installed a 360-horsepower fuel-injected 327 in a Chevy II station wagon and promptly won the B Factory Experimental Class at the 1962 NHRA Winternationals. Nicholson's quarter-mile times were in the mid-12s at close to 110 miles per hour, bettering the times run by many big-cubic-inch super stockers.

Chevrolet

The body structure for the 1962 Chevy was the same as the one used in 1961. The passenger doors and a majority of the uppers were a direct carryover.

The two-door sedans had a new roof, eliminating the flat-top design of the 1961s. The 1962 Bel Air Sport Coupe retained the fastback roofline of the 1961. The Impala Sport Coupe was given a new, more formal, cabriolet look with a roof that resembled a convertible top with creases across the rear portion. The side moldings were thinner and horizontal, giving the car a longer look. The rear was made to look wider by virtue of a large anodized aluminum panel, and the grille was a simple egg-crate design. The front and rear bumpers had a much larger section, eliminating the painted valence panels of the 1961s. The front wheel houses were now a full skirt design, protecting the inside of the fender from road spray and corrosion.

Each of the three Chevrolet models, Biscayne, Bel Air, and Impala, had its own interior trim levels. The Biscayne was the most Spartan with rubber mats for a floor covering. The Bel Air had full carpeting and upgraded seat trim. The Impala interior was the most plush, featuring anodized aluminum trim on the instrument panel. The 1962 Impala seats, like those on the 1961s, had a design pad sewn in. This time it took the shape of buttons and biscuits in the seatbacks. Ed Cole liked to spend the money where it could be seen.

For 1962, the Impala Super Sport option was reformulated. Customers were no longer required to purchase a long list of options to get the sporty SS package. All that was required was the purchase of a base Impala Sport Coupe or convertible. There were no engine requirements. The 1962 SS could even be ordered with a six-cylinder engine. This brought the price point down to a reasonable level of $145 over the base Impala. A 1962 Impala Sport Coupe with the base 283 V-8 and the Super Sport option listed for approximately $2,680, a far cry from the 1961 SS price of nearly $4,000. This new lower price and positive exposure of the 1961 SS attracted many customers. For $145 the customer got a passenger-assist bar and an all-vinyl bucket seat interior. The bucket seats were Corvair units, since the Chevrolet Division could not afford the extra expense of using the more-luxurious Strato bucket seats installed on Pontiacs. Between the seats was a small locking console. Four-speed Super Sport Impalas had a chrome shifter plate similar to the one installed on the 1961 SS. The exterior of the 1962 SS featured tri-bar knock-off spinners on the center of the 1962 full wheel cover. Thin-line white sidewall tires, optional only on the 1961 SS, were the only optional white sidewall tires available for any 1962 Chevrolets. The full-length body side molding insert and rear taillight panel were engine-turned. Special SS quarter-panel emblems finished the exterior. Chevrolet sold just under 100,000 1962 Super Sport Impalas.

To celebrate its golden anniversary in 1962, Chevrolet changed the color of the small bow tie insert on the hood and deck lid emblem from blue to gold. Many dealers featured fully loaded Impalas painted Autumn Gold to draw traffic into their showrooms. New car customers in 1962 were given shiny brass key chains that said, "Thank you, America, for 50 years of confidence." Gimmicks were hardly needed, though, as the 1962 Chevys sold well.

The 1962 Chevrolet buyer was presented with exciting new styling and a selection of six engines from which to chose. Anchoring the line-up was the 235-cubic-inch 135-horsepower six-cylinder engine. A new partial-flow oil filter was standard equipment on the reliable six. The base V-8 was the 170-horsepower 283. It was equipped with a two-barrel carburetor and single exhaust. Both engines were backed by a column-shifted three-speed manual transmission. Optional was a three-speed overdrive and a Powerglide.

New for 1962 were two 327-cubic-inch engines, rated at 250 and 300 horsepower, replacing the 348-cubic-inch W engines. The 327 was based on the 283 small-block design. Chevrolet engineering experimented with boring and stroking the 283, and found that a 0.125-inch bore and a 0.250-inch stroke produced an engine with 327 cubic inches. This new combination generated as much power as the 348, was more fuel-efficient, and was 115 pounds lighter.

From the outside, the 327 was the same package size as the 283. New blocks were cast to accommodate the increased bore and stroke. Extra material was added to the main bearing webs for strength, and the area around the bores was increased. The 327's crankshaft was forged and required machining of the block for counterweight clearance. The rods were the same as those used in the 283, except they were beefed-up in the beam section. Pistons were cast with flat tops with machined reliefs for valve clearance. The compression ratio was 10.5:1, necessitating premium fuel.

Cylinder heads for the 250-horsepower 327 were basic 283 units. The 300-horsepower version used the 1961 Corvette's 315-horsepower heads with their larger ports and valves. Small four-barrel carburetors manufactured by both Carter and Rochester were used with the 250-horsepower engine, while the 300-horsepower 327 used the larger Carter AFB. Distributors for both engines were single-point vacuum-advance. Both engines

The 1962 Corvette was a mildly updated version of the 1961 Corvette. A new rocker panel molding was added and the coves were no longer available in a contrasting color. The big news for the 1962 Corvette was the selection of new 327-cubic-inch engines ranging from 250 horsepower to a fuel-injected version rated at 360 horsepower. *GM Media Archives*

came standard with dual exhaust and a temperature-controlled fan. Three-speed syncromesh transmission was standard and the aluminum Powerglide and four-speed manual were optional.

Topping the horsepower chart for the 1962 Chevrolets were two 409s, a single four-barrel version rated at 380 horsepower and a dual-quad version rated at 409 horsepower. Both horsepower ratings came as a result of the service package released late in the 1961 model year and installed on production 1962 409s. That package included head castings with larger ports and valves, new intake manifolds, and a new longer-duration camshaft. In April 1962, another service package was released for the 409 containing a new cam, a single-point vacuum-advance distributor, and new streamlined exhaust manifolds. With this service package the advertised horsepower ratings for the two 409s did not change. All 409s came from the factory with dual head gaskets, which dropped the advertised compression ratio of 11.0:1 about one point and made the 409 a bit more driveable on the street.

Semon "Bunkie" Knudsen

Semon "Bunkie" Knudsen was born on October 2, 1912. He was the only son born to William S. Knudsen, former general manager of Chevrolet Division and president of General Motors. Father and son were very close. The nickname Bunkie—short for "bunk mates," or good buddies—was given to him by his father.

As a young lad, Knudsen had an interest in cars, and when Bunkie turned 14 his father responded to his request for a car with an automobile in hundreds of pieces. Before he could drive it, he had to put it together, and he did.

Bunkie graduated from the Massachusetts Institute of Technology in 1936. In the three years following graduation, he worked in several Detroit-area machine shops, and in 1939, he joined General Motors on the manufacturing staff of the Pontiac Motor Division. In 1956, Knudsen was promoted to general manager of Pontiac Division, and he worked magic on what was a sedate car line. His axiom, "You can sell a young man's car to an old man, but you can't sell an old man's car to a young man," became the anthem for Pontiac. The Pontiac name stayed the same, but the image changed dramatically. Wide-track chassis, aggressive new styling, tri-power carbs, and four-speed transmissions were added, and Pontiac's sales soared.

In 1961, Knudsen followed Ed Cole as general manager of Chevrolet Division. During the four years he headed the division, two new nameplates were born (Chevelle and Chevy II) and the Corvette was completely redesigned. Knudsen brought to Chevrolet the same excitement he brought to Pontiac, setting new sales records each year. Knudsen's team redesigned the 1965 Chevrolet and introduced the 396SS Chevelle.

It was as if Pontiac was the training ground for future Chevrolet general managers. Following Knudsen as head of Chevrolet were Pete Estes and John De Lorean, and the trio was affectionately known as "The Three Musketeers."

Knudsen wanted something new to talk about every year. He was loved by the designers because he

S. E. "Bunkie" Knudsen

encouraged good design, applauded it, and supported it. He didn't make the designers think about costs; he freed their creativity. When Knudsen walked into the studio, he got excited about what the designers were working on and saw the possibilities at the showroom level. The designers worked hard for Knudsen because he appreciated what they did. He, like Ed Cole, fought GM's Board of Directors for the exciting products Chevrolet's engineers and designers were creating.

Knudsen loved high-performance cars and was a big fan of NASCAR. On the streets of Detroit he often drove a Biscayne sedan. Under the hood lurked the strongest big-block that Chevrolet engineering could build. In the early 1960s, many a young Turk had his doors blown off on Woodward Avenue by this middle-aged man in a plain-Jane Chevy sedan.

Chevrolet figured that if someone were to seriously race a 409, they would tear it down and rebuild it with a single head gasket.

Powerglide automatic was not available with the 409. Both 409s came standard with a column-shifted three-speed manual transmission, but very few were produced with that transmission. Most 409s were ordered with the optional close-ratio four-speed manual transmission. With that four-

speed came a column-mounted Sun tachometer that red-lined at 6,200. Sales literature said the 409 was available in any body style. However, dealers refused to order high-performance cars with unusual combinations (like a station wagon with a 409 and a three-speed) to keep from being stuck with a car they couldn't sell.

All 1962 Chevys, including the 409s, were covered by a standard manufacturer's warranty. Dealers soon got wise to young drag racers and their penchant for parts destruction. It was not unusual for the salesman to take the youthful 409 owner aside and tell him, in a fatherly way, that all he would get out of the warranty was one engine, one clutch, and one transmission—and then he's on his own.

Chevrolet had such success with the 409 in 1961 that by the time the 1962s were released, Chevrolet's Los Angeles zone office had back orders for more than 500 409s. The 1962 409s picked up where the 1961s left off—winning races. "Dyno" Don Nicholson again swept the NHRA Winternationals, driving an Ermine White 1962 Bel Air Sport Coupe powered by a dual-quad 409. This time the competition was much stiffer from Ford, Mopar, Pontiac, and other Chevrolet camps. At 12.84 seconds, Nicholson's times were a half second faster than his 1961 winning ET.

Nicholson knew how to make a 409 run and was one of the best drivers in the business. Chevrolet never formally backed Nicholson, but occasionally gave him transmissions and rear ends when they broke. They never had to give him an engine, however, because he never blew up a 409. Dual-quad 409s spent the spring and summer mopping up the competition in match races. When the NHRA U.S. Nationals were run in September, the winner was once again a 409, this time driven by Hayden Proffitt. Unfortunately the 409s did not fair as well on the NASCAR circuit. They were reliable, but didn't have the power in the longer runs to match the 421 Pontiacs.

Chevrolet was riding the crest of the wave in 1962. The restyled body and new roofline were well-accepted. Every enthusiast magazine that tested a Chevrolet in 1962 raved about it. Chevy was definitely on a roll.

Corvette

The 1962 Corvette was, with a few exceptions, very similar in appearance to the 1961 model. The side-cove color option was no longer available and a chrome rocker molding was added. New crossed-flag emblems were on the sides of the front fenders and on the area in front of the hood. Optional white sidewall tires were now thin lines.

The big news for Corvette was to be found under the fiberglass hood. Owners of 1962 Corvettes were the beneficiaries of the new 327-cubic-inch engine originally designed to replace the 348s in the full-size passenger car. The 250-horsepower model, which was the base engine for the Corvette, and the optional 300-horsepower version were identical to the engines installed in the full-size Chevys. Corvette owners also had two high-performance versions of

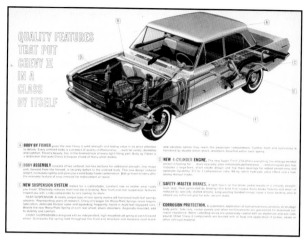

This cutaway drawing and accompanying descriptive text were taken from the 1962 Chevy II sales brochure. They detail the engineering features built into Chevy's newest car. One feature not mentioned is that the engine compartment was designed to accommodate the small-block V-8—an option that would not be available until 1964. In 1962, V-8 installation kits were available from the Chevrolet parts counter.

the 327 from which to choose, rated at 340 and 360 horsepower. Both engines had forged 11.25:1 pistons, a solid lifter camshaft, and free-breathing high-performance Corvette heads from 1961. Like the 409s, both high-compression 327s had dual head gaskets installed at the factory. All distributors on 1962 Corvettes had a mechanical tachometer drive. The 250 and 300 engines had single points with a vacuum advance, and the 340 and 360 had dual points with no vacuum advance. The 340-horsepower engine had a single Carter AFB carburetor on an aluminum intake manifold. The carburetor was similar to the one on the 300-horsepower engine. The 360 engine had Rochester fuel injection, which was revised in 1962 for quicker starting and improved cold weather operation.

A floorshifted three-speed manual was the standard transmission, with Powerglide optional only for the 250- and 300-horsepower engines. A four-speed was also optional with wide-ratio gearing for the two smaller engines and close-ratio gearing for the two solid-lifter engines. These ratios could be swapped on special order.

Standard brakes were 11-inch drum-type brakes, while sintered metallic linings were optional along with a second, RPO 687 roadrace chassis. This special option increased the steering ratio from 21:1 to 16.3:1 and added heavy-duty shocks. Special finned brake drums with added area and sintered metallic linings were used. Air scoops were used on the backing plates to aid in cooling, along with stamped steel fans on the face of the brake drum to increase air circulation. RPO 687 was available only with the 360-horsepower fuel-injected engine and positraction rear axle. Another option created for roadracers was a 24-gallon gas tank, which was only available with the hardtop, since it used up the space normally reserved for the folded soft top.

Rumors were circulating about the 1963 Corvette. It was difficult to imagine how Chevrolet could produce a car that would capture the hearts of performance enthusiasts any more than the 1962 Corvette had.

Corvair

In the spring of 1962, quite a few changes came about for the Corvair. The slow-selling station wagon was dropped. The Corvair wagon was as much competition to the new, more-spacious Chevy II wagon as it was to competitors' station wagons. The new Monza convertible was introduced as part of the Monza series, and it fit well into the mini-sports car niche the Corvair was now filling.

The most exciting news for Corvair enthusiasts was the April 1962 release of the new Spyder option, RPO 690. The core of the Spyder was its turbocharged engine. Chevrolet wanted to offer the first modern car with a turbocharger, but the Olds F-85 beat it by a few months. Under the direction of Robert Benzinger, Chevrolet engineers James Brafford and Robert Thoreson worked out the turbo installation. During their investigation, they explored the possibility of a mechanical supercharger, but it was soon eliminated from consideration because of the horsepower it drained and the already unique fan belt installation. A turbo was favored because it freewheeled at cruise speed, and installation was much simpler since the exhaust gas-driven turbo did not need a belt or gears to drive it. It also solved the carburetion problem, since one side-draft carburetor replaced multiple carbs.

A basic 102-horsepower Corvair engine, with a few modifications, was used for the Spyder. The compression was dropped from 9:1 to 8:1. A stronger crankcase, crankshaft, and rods were part of the extensive internal modifications. The Spyder had a unique spark advance curve. Initial timing was set at 24 degrees, reaching a maximum of 36 degrees at 4,500 rpm. Maximum boost was 11 psi at 4,000 rpm, at which point it dropped to 10 psi. Manifold pressure was used to retard the spark a maximum of 9 degrees, and premium fuel was required.

Upon the release of the Spyder option, Chevrolet required the purchaser to also buy the four-speed

Corvair Monza Club Coupes were hot sellers in 1962. At 129,544 units, they represented almost half of all Corvairs sold in 1962. This Monza Club Coupe stickered for $2,273 and included bucket seats, bright window trim, rocker panel molding, full wheel covers, and special Monza nameplates. *GM Media Archives*

transmission, heavy-duty suspension, and metallic brake linings. Positraction and seatbelts were recommended but not required.

Below 3,000 rpm there was a noticeable lag in acceleration, but above 3,000 it took off like a rocket. Zero-to-60 elapsed times for the Spyder were as low as 10 seconds flat, compared to 15.5 seconds for the standard 102-horsepower engine on which the Spyder was based. Top speed rose by almost 20 miles per hour to 110. With this new engine, the Corvair had finally earned its sports car "wings."

The direction of all Chevrolet products was now clear—*performance*. The new 327-cubic-inch small-block came out of the box as a reliable performer and bolstered full-size car sales. The 409s quickly vaulted to legendary status by virtue of their drag strip exploits. The new Chevy II, introduced as an economy car, could be upgraded easily to V-8 power with a kit available through the dealer. The new 327 increased the performance of the Corvette without any weight penalty, and the Corvair upped its rank in the performance arena with the addition of the turbocharged Spyder option. The engineering department at Chevrolet was certainly busy!

All of this horsepower activity brought customers with money into the showroom. In 1962, Chevrolet sold 2.1 million passenger cars, almost one-third of all cars sold in the United States. Chevrolet far outstripped Ford's 1962 sales of fewer than 1.6 million. In 1962, Ford openly said it no longer supported the AMA's 1957 ban on horsepower or racing and actively supported racers using their products. Chevrolet supported the ban in principle, but supplied resources to racers through the back door while the engineering development went forward at a torrid pace.

Chapter 4

1963 A New Corvette and an End to Racing

In the early 1960s, Chevrolet and General Motors strove to build better cars. They poured a great deal of money into engineering and testing to create a better product. GM wanted to be confident each car had a balance of performance, reliability, durability, safety, comfort, appearance, and economy.

Most of the vehicle tests were conducted at the GM Proving Grounds in Milford, Michigan, a facility that opened in 1924 with 70 miles of every conceivable type of road. Testing for GM's 1963 models ran up a total of 20 million miles, as each GM product line, including Chevrolets, was given a 36,000-mile durability run. Following this test, the cars were disassembled and each part was inspected for wear. In 1963, GM added to the Milford facility an impact sled to simulate auto crash-worthiness. Additional road test facilities were built in Mesa, Arizona, and Pikes Peak, Colorado, to permit testing of vehicles in extreme heat and at high altitude. General Motors felt confident its cars were the best value available, and to back up that claim, it extended the new car warranty for all 1963 GM cars to 24 months or 24,000 miles.

Several new features were standard on all new 1963 Chevys. Self-adjusting brakes were a welcome

This Chevrolet studio photo depicts a Riverside Red 1963 Corvette convertible against the backdrop of a futuristic monorail. The fuel injection emblems on the side of the front fender indicate a 360-horsepower 327-cubic-inch engine. Engines and transmissions were the only items carried over for 1963's new Corvette. Sales in 1963 were almost equally split between coupes and convertibles.

When the 1963 Corvette was designed, the coupe body style was the first to be created. Using a totally unique approach, the window and door cut lines were scribed on the surface after the body was styled. The split rear window on the 1963 Sting Ray coupe was the subject of much controversy within GM. Mitchell insisted on it; Duntov hated it.

safety addition, as no longer did the owner have to pay to have his brakes adjusted every 5,000 miles. A simple linkage installed on each brake adjusted the shoes each time the brakes were applied when in reverse. Delcotron alternators became standard on all passenger cars, except the Corvair. The Delcotron and its regulator were smaller and lighter than the old generators and provided a better charge at low engine speeds.

In 1958, an industrywide project was launched to evaluate all motor vehicle lighting. It was determined that amber-colored turn signal lights could be seen more easily against the glare of white headlights. In 1963, under the recommendation of the AMA, all Chevrolets were equipped with amber front turn signals. PCV valves, optional in 1962, were standard on all 1963 Chevrolet passenger cars. This simple device reduced air pollution and extended engine life by ridding the crankcase of harmful gases. Finally, corrosion-resistant mufflers and exhaust pipes were added to all GM car lines. These additions, included in a sticker price not much higher than that of 1959, proved Chevrolet was the value leader in 1963.

Bunkie Knudsen always wanted something new and exciting to talk about with each new car introduction, and in 1963 he had a lot to talk about. Corvette's new knock-out styling and chassis design carved a deep niche in automotive history.

Chevrolet car sales in 1963 totaled 2.2 million, far outstripping the nearest competitor, Ford, which sold just over 1.5 million units. General Motors' total output was 9 percent above 1962 levels, with just over 4 million units sold—well over half the U.S. auto production of 1963.

Corvette

The 1963 Corvette carved a niche in automotive history equaled by few other cars. Bill Mitchell was the true architect of the 1963 Corvette, with young designer Larry Shinoda interpreting Mitchell's vision and putting it on paper. The 1963 Sting Ray's classic lines were rooted in Mitchell's famous Sting Ray race car of the late 1950s. Mitchell wanted a car that was special. He desired one smooth shape with a body character line that went all the way around the body. Over the wheel openings, the body was pushed up to make room for the tires. One of Bill Mitchell's favorite shapes was the boat-tail contour. This form was applied to the coupe's fastback roofline that tapered to a point in the rear. Mitchell made it clear right from the beginning that Chevrolet was going to build both a coupe and a convertible.

The initial design for the 1963 production Corvette was the coupe. The roadster came later by simply removing the roof. Once the total shape of the coupe body was designed, the stylists then added the door and window cut lines. The new coupe was radically different, having no quarter windows. A stir was created within Chevrolet over the split rear window. Corvette engineers, headed by Zora Arkus-Duntov, hated it. They felt this styling whim severely limited rear vision. Mitchell insisted the split continue through the rear window. The Chevrolet stylists loved it. It was as if a stingray were floating on the roof, its whip-like tail trailing down the rear window. It was a classic internal corporate battle, one which Duntov lost and Mitchell won. Mitchell also wanted scoops in the rear fenders. Production designers removed them in favor of simulated gill vents on the sail panels. These fake gill vents became functional on 1964 coupes.

Hidden under the new Corvette body was a network of steel structural members. It acted as a backbone to increase the torsional rigidity of the Corvette's fiberglass body. Structural body sill members were tied to both lock and hinge pillar reinforcements. Hinge pillars were tied together laterally with cowl and plenum members, forming a strong bridge across the car. The coupe's passenger compartment was completely framed in steel with a member running across the roof between the lock pillars. This steel structure increased the Corvette's weight, but the benefits far outweighed the small weight disadvantage.

An all-new interior complemented the new Corvette's beautiful exterior. One complaint registered about the previous Corvettes was the lack of a comfortable seating position. Steering wheel placement was too upright and close to the driver. Legroom was at a minimum. Most first-time drivers of the earlier Corvettes found the seating difficult to get used to. Seating in the new Corvette was much more comfortable. The driver sat more "in the car" than "on the car." A full 3 inches of fore and aft adjustment was designed into the new Corvette's steering column. Topping off the steering gear was an attractive new deep-dish, three-spoke wheel. For the first time in Corvette history, power steering was offered as an option.

Driver and passenger both sat in vinyl-covered bucket seats. Optional saddle-colored leather seats were available for $75. Door panels were trimmed to match the seat and each had a small armrest. The vinyl-covered instrument panel featured two hooded coves. Instrumentation directly in front of the driver included two large circular dials for the tach and speedometer. They were surrounded by four smaller dials for gas level, amps, oil pressure, and water temperature. A locking glovebox was positioned within the passenger's cove area, and a passenger assist bar was cut into the top of the instrument panel, while an electric clock and AM radio filled the area between the two coves. AM/FM radio was a midyear addition to the option list.

Chuck Jordan, head of GM's exterior design from 1962 through 1967 and eventual GM vice president of design, couldn't believe Chevrolet was actually going to produce a car like the 1963 Corvette. "Development of the 1963 Corvette was a profound moment in the history of Chevrolet," said Jordan. "I can't remember any time where I felt as enthusiastic and exuberant about a car. There were no compromises on the '63 Corvette—sometimes things just fall together."

Jordan also felt the convertible, which came about following the coupe's design, lost a lot of identity

without the coupe's unique upper body. Jordan's personal car in 1963? A silver split-window coupe.

Some of the initial proposals for the 1963 Corvette included a four-passenger hatchback coupe. Larry Shinoda called the four-seat Corvette "Ed Cole's brilliant, but bad idea." Cole hosted a design review comparing a 1961 Thunderbird to the proposed four-passenger 1963 Corvette. During the review, then-GM President John Gordon asked Cole if anyone could really fit comfortably into the back seats. Cole assured him of the comfort level, opened the right side door, and pulled the seat forward for Gordon to enter. Gordon got into the back seat, but didn't fit too well and wanted out. The seat wouldn't release, thus trapping the GM president in the back.

The space was small and it was a warm, humid day in Detroit. Someone ran to get a wrench to unbolt the seat and emancipate the overheated and now irate executive. With sweat soaking through his expensive suit, John Gordon emerged from what was the first and last four-passenger Corvette.

"That killed that program and everybody was relieved," said Shinoda. "It looked terrible, it had a humped-out roof and a real long door. It didn't make sense." There was one benefit to the four-passenger design: a stronger frame. The frame rail section had been designed for the weight of a four-place vehicle. When the decision was made to go back to a two-passenger vehicle, the section size of the frame was preserved.

46

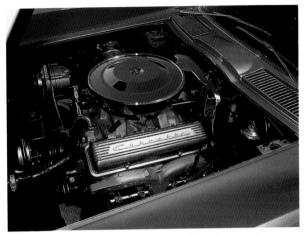

All 1963 Corvettes were powered by one of four versions of Chevy's 327-cubic-inch engine. Horsepower ratings ranged from 250 to 360. This particular engine was rated at 340 horsepower and was equipped with a solid-lifter cam and single four-barrel carburetor on a high-rise aluminum intake manifold.

The 1963 Sting Ray was probably the most revolutionary car Chevrolet produced in the 1960s. Beneath its beautifully inspired body was a chassis worthy of a world-class sports car. Under the hood, Chevrolet's 327-cubic-inch V-8 could be ordered in horsepower ratings as high as 360, making the new Sting Ray as fast as any car on the road.

Attached to that sturdy frame was the most modern suspension ever put under an American passenger car. Chevrolet passenger-car components made up the majority of the coil spring front suspension that included a sway bar. Corvette steering gear was all new for 1963 and was based on the passenger car design. A steering ratio change could be easily made on the new Corvette to improve steering response. Each steering arm had two holes for the tie rod socket, and normally, the hole at the end of the arm was used for the standard ratio. All that was needed to switch to quick-ratio steering was to move the tie rod into the forward hole. Vehicles equipped with optional power steering had only one hole on the steering arm and could not be modified to change the steering ratio.

Prior to the release of the new Corvette, most American automotive enthusiasts had never heard of an independent rear suspension. But in 1963 it became a household word. Corvette's rear suspension was all new in design and componentry. It featured a transverse-mounted multileaf spring and a fixed differential carrier. Each rear wheel was held in place by a half shaft, a strut rod, and a trailing arm. The half shaft and strut rod combination provided parallel arms that controlled the rear suspension's vertical travel. Movement fore-and-aft was limited by the trailing arm, which also mounted the rear brakes. Ring and pinion ratios were the same as in 1962 with positraction as an option. The 1963 Corvette's new rear suspension was simple, strong, and effective.

With the exception of a few revisions on the fuel injection unit and air cleaners, the 1963 Corvette's engines were identical to those offered on the 1962 models: four 327s ranging in horsepower from 250 to 360. Finned aluminum rocker covers were standard on the two high-performance engines; painted,

Veteran drag racer Don Nicholson (third from left) was one of 57 racers who had the privilege of paying an extra $1,237.40 for the Z-11 option on his 1963 Impala. In addition to the special 427-cubic-inch engine, the Z-11 package featured an aluminum hood, fenders, and bumpers, which trimmed 300 pounds off the Impala's weight. It was the only car Chevrolet ever built specifically for drag racing. *James J. Genat*

stamped-steel was standard on the two low-end 327s. Low-profile chrome air cleaners were standard on the three carbureted engines, and Rochester's fuel-injection unit had a much lower profile than those of previous years as its revised manifold was cast without internal bulkheads and had a removable cover.

Transmissions for the 1963 Corvette were the same as those offered in 1962: a three-speed, two four-speeds (close- and wide-ratio), and Powerglide. The aluminum bellhousing was the same one found on the passenger car.

Fuel capacity was increased from 16.4 gallons (1962) to 20.0 gallons (1963). The fuel filler was in the center of the rear deck under a flip-up emblem. An optional 36.5-gallon fuel tank was part of special performance equipment option Z06, available only on coupes. Later in the year, the big tank was available as an individual option, again only on coupes.

Magazine editors were anxious to get their hands on the new Sting Rays, and Chevrolet was more than willing to loan a few out. Reviews were marvelous! After studying and driving the new Corvette, Ray Brock wrote in the October 1962 issue of *Hot Rod* magazine, "This car is sensational in every way. Styling acceptance is a matter of personal taste but we know it will be a hit. Handling is superb. Brakes are great. You can order as much power as you want. . . . Everything we ever liked about a Corvette has been vastly improved and the little things we didn't like have been eliminated."

Chevrolet designers were also eager to get behind the wheel of the new Corvette. Designer Dave Holls had the chance to drive a silver Sting Ray coupe two days before they were released. "I headed for Ted's Drive-In on Woodward Avenue," said

Corvair's Spyder option (with its special 150-horsepower turbocharged engine) was first released in 1962 and available only in the Monza coupe. In 1963, Chevrolet expanded the availability to include the Monza convertible. The Spyder option stickered for $317.45 and, in addition to the 150-horsepower engine, included special instrumentation and exterior emblems. In 1963, just over 19,000 Corvairs were sold with the Spyder option.

In January 1963, General Motors announced its decision to abide by the 1957 edict of the AMA and stay out of racing. In February, five Impalas equipped with Chevrolet's new "Mystery" Mark II 427 rolled into Daytona for the annual 500-mile race. The Mark II 427 engine had been kept a secret from all but a few people within Chevrolet. Junior Johnson poses in front of his 427-powered Impala that qualified at 163.68 miles per hour—10 miles per hour quicker than the fastest Chevy in 1962. *International Speedway Corporation/NASCAR*

Holls. "I pulled in with that coupe and every car in the place emptied to get a closer look."

Up until 1963, Corvettes were predominately bought by men. It's not that women didn't want them, it's just that Corvettes lacked the amenities needed to soften their hard-edged sports car image. With the additions to the Corvette option list of power steering and power brakes, that edge was finally softened. The Corvette now appealed to a larger market without hurting its performance image, as sales figures confirmed.

Chevrolet

The newly restyled 1963 Chevrolet once again had the look of a mini-Cadillac. Out of sight, beneath the new exterior, was the 1962 chassis. All the rooflines were carryovers, with the exception of the newly styled four-door hardtop, and all the sheet metal below the beltline was new. Two pronounced horizontal character lines ran along the sides, which gave the Chevy an illusion of length. Thin horizontal side trim ran along the lower of those lines, enhancing the aspect of length. All 1963

bodies received a new, straight A-pillar design, an innovation that reduced wind noise and provided a better seal for the windshield. The rocker panels were designed to allow fresh air from the cowl to flow through them, helping to flush out trapped water, thereby reducing corrosion.

An enlarged grille opening gave the cars a wider, more massive appearance. It was the traditional Chevrolet egg-crate design executed in anodized aluminum. The rear of the new Chevy Impala was coved out with six circular taillights placed on an aluminum panel. All other models had only four taillights outlined with a thin aluminum molding.

While the 1963 Chevys were being designed, chief designer Clare MacKichan was recuperating from a bad car accident. His assistant Dave Holls, who had just transferred in from Cadillac, was in charge of the studio. Every evening Holls went to MacKichan's house and reviewed what was being done. "The '63 Impala had a lot of Cadillac in it with that low color-accented body molding, and the front looked a little like a '61 Caddy," said Holls. "Because of the similarity, they used to kid me that the leopard didn't change its spots."

The 1963 Chevrolet interiors also received a new look. Designers again challenged the production folks to see how far they could bend a piece of sheet metal for the new instrument panel, a very smooth, sweeping design that housed the instruments in a deep recess. The glovebox was moved back to the right-hand side in front of the passenger seat. Even though it was still an add-on, the tachometer was better integrated into the instrument panel design and was located directly in front of the driver at the base of the steering column, housed in a slick nacelle with a bird beak at the top. The tach was a Delco Electronics product, standard with any four-speed 409 or 300-horsepower 327 and optional on all other V-8s. Seats in all models received new materials, and the Impala carried the "button and biscuit" pattern design pad in the seatbacks.

The 1963 Super Sport featured vinyl-covered bucket seats trimmed in the same pattern as the Impala. A new, larger console with a floorshift split the buckets when a Powerglide or four-speed transmission was ordered. A passenger assist bar was not part of the 1963 SS trim. Super Sport exteriors featured the SS badges and engine-turned finish for both the side molding and rear insert, and special three-bar spinner wheel covers rounded out the SS exterior trim. As in 1962, the SS option could be ordered with any engine, but only in an Impala sport coupe or convertible body style.

Several new production options appeared for the 1963 Chevrolet. For $70, a vinyl top covering, in white or black, could be ordered on the Impala Sport Coupe. A new seven-position tilt steering wheel was available on all Bel Air and Impala series with either a Powerglide or four-speed manual transmission. For the first time Chevrolet offered an AM/FM push-button radio as an option. At $125, the AM/FM radio was almost double the cost of the AM push-button radio.

Small-block V-8 engines for 1963 were carry-overs. Now there were three 409s to chose from and they were all trimmed in chrome. Horsepower ratings for the two solid-lifter 409s were increased to 400 and 425. While rated higher, these engines were the same mechanical configuration as those offered in the service package available in April 1962. Added to the 409 line-up for 1963 was a tamer 340-horsepower version for the street. The 340-horsepower 409 had cylinder heads with smaller ports and a cast-iron intake with a small four-barrel carburetor. The 340 had a mild hydraulic cam and a 10:1 compression ratio that required premium fuel. This mildest of 409s came standard with a three-speed

This is where Chevrolet executives expected to see their new Impala—sitting freshly washed in the driveway of a middle-class suburban home. All 1963 Impalas came standard with small hubcaps, and full wheel covers were optional. Due to the wide 6.0-inch wheels (standard with the 409 engine), the small hubcaps on this Impala appear to be recessed into the wheel. *James J. Genat*

manual transmission. Optional was a wide-ratio four-speed or a heavy-duty Powerglide. The 340-horsepower 409 had a beefy 420 foot-pounds of torque at 3,200 rpm. *Car Life* magazine tested a 1963 Super Sport with a 340-horsepower 409 backed by a Powerglide. It took only 6.6 seconds for the Impala to reach 60 miles per hour from a standing start.

At the low end of the Chevrolet horsepower spectrum was an all new six-cylinder engine. This engine was long overdue, as Chevy had been using a six designed in the 1930s. The new 230-cubic-inch six was based on the design of the Chevy II's six, but with a larger bore. The new six was 2 inches shorter and 3 inches lower than the previous six, and this compact size reduced the weight by 23 percent. Many parts on the new six are interchangeable with the four and six found in the Chevy II.

Late in 1962, Chevrolet's clandestine support of organized racing was increasing. Still not officially condoned by management, extensive high-performance programs were going forward at flank speed behind closed doors. Chevrolet wanted to maintain its dominance in drag racing and start winning on NASCAR tracks. To accomplish those goals, Chevy launched an all-out-assault in both arenas.

New rules from both NHRA and NASCAR were put in place in an effort to rein in the factories. Engines on NHRA "stock" cars were limited to a maximum of 427 cubic inches. This rule followed the 427-cubic-inch maximum set by NASCAR to limit what was foreseen as an imminent cubic-inch war between the Big Three. Minimums were also set as to the numbers of cars produced on the assembly line that could be legal for stock classes. This was intended to discourage the production of "one-off" specialty race cars.

Two special engines were installed in 1963 Impalas, one exclusively for drag racing and one exclusively for oval tracks. Performance of both engines sent the competition scrambling back to their drawing boards to come up with more horsepower. NASCAR got a brand new engine for 1963 dubbed the Mystery 427, while NHRA got the Z-11 package.

The 1963 racing season dawned with Chevrolet ready to go with the Z-11-optioned Impala. The Z option designated a regular production option (RPO) which included a special body, chassis, and engine combination. The Z-11 was a full-combat model—lightweight and high horsepower. Fifty-seven production Z-11 Impalas were built between December 1962 and January 1963, at the Flint, Michigan, assembly plant. Immediately, a few high-profile racers like Don Nicholson, Bill Jenkins, and Malcolm Durham, received their cars from Chevrolet Division. Other Z-11s were dispatched to dealers to be sold to quarter-mile hopefuls.

Conservatively rated at 430 horsepower, the Z-11's 427 engine was the final iteration of the W engine. With the exception of the large black air cleaner, the engine looked like a production 409. It was painted Chevrolet-Engine Orange with silver rocker covers. There was no valve cover identification as to cubic inches or horsepower.

The Z-11's 16 extra cubic inches came from the addition of 0.150-inch in stroke to the crankshaft, which was installed with no block modifications. Forged pistons provided the 13.5:1 compression. The Z-11's solid lifter cam had 325 degrees of duration and 98.5 degrees of overlap, and its water pump was made of a lightweight aluminum casting.

The major difference between the 409 and Z-11 427 was the cylinder head design. Z-11 heads were unique and could only be used with the Z-11 intake/valley cover. Z-11 rocker studs were a screw-in design, unlike the pressed-in and pinned studs of the 409. Unique to the Z-11 was the pushrod guide plate, which bolted to the head under the valve cover. Z-11 valve covers were similar to the 409 covers with one small exception: front and rear corners on the edge closest to the intake were squared off for clearance for the pushrod guide plates.

The Z-11 had a high-rise cast-aluminum two-piece dual four-barrel intake manifold. This two-piece design separated the intake manifold from the hot engine oil, a design which allowed air to circulate beneath the manifold, cooling the intake

The Monza was at the top of the 1963 Corvair series. The Monza was offered as either a two-door coupe, a convertible, or a four-door sedan. The coupe and sedan had bright trim around the door frames and a stainless drip rail cap. Behind the front wheel on the fender of all Monzas were special Monza nameplates. *GM Media Archives*

charge. It mounted the same Carter AFBs as the 1963 425-horsepower 409.

The Z-11 air cleaner was an engineering work of art. Its simple design took advantage of the high-pressure area at the base of the windshield. On a passenger car, the air trapped in this area is channeled into the cowl, providing fresh outside air for the heater and interior vents. For the Z-11, Chevrolet simply cut an elongated hole through the firewall into the cowl plenum and added a flange that connected to the air cleaner assembly. Unfortunately, most racers discarded this well-engineered assembly for the more traditional hood scoop.

Suspension on the Z-11 was as unique as every other aspect of the car. Heavy-duty shocks and springs were fitted all around. Drum brakes had sintered metallic linings without power assist. Each front brake backing plate had large cut-outs covered with a metal screen. These vents helped cool the brakes after each 120-mile-per-hour quarter-mile pass. All V-8 Chevrolets in 1963 were equipped with a standard front sway bar, except the Z-11, and deletion of this bar and its associated hardware reduced front-end weight. There was no real need for a sway bar, since the only turns this car would make would be at low speed onto the drag strip's return road.

The rear suspension had an additional upper control arm. Standard Chevrolet suspension had two lower arms with a single upper arm on the right side and a panhard bar. The additional upper control arm's bracket was welded to the left side of the differential housing. Its upper end mounted to the same frame cross-member as the right side upper arm. This upper arm modification reduced pivoting of the rear axle. It was something racers had learned in 1962, and it was applied by the factory to the Z-11 cars. The

Custom Accessories

In 1963, many new Chevrolet options were not installed on the assembly line. In fact, most custom accessories were installed at the dealer prior to delivery to the customer. These dealer-installed items increased the dealer's profits on every new car sale. Most common among dealer-installed accessories were clear plastic seat covers. These were installed to protect the factory cloth from wear-and-tear or from the dirty work clothes of a blue collar owner.

Chevrolet designed and manufactured many custom accessories exclusively for its dealers to install. Many of the dealer-installed items, such as power brakes, temperature-controlled fan, or full wheel covers, were available as regular production options. Full wheel covers ordered on a new Impala or Bel Air in 1963 cost $17.00; if purchased at the dealer, they listed for $24.50. Dealers frequently negotiated the price with customers, since margins on these items were already high.

The list of items available only as dealer-installed options was lengthy. For $23.00, plus installation, the owner of a new Chevy could have a rear window defroster. It was nothing more than a blower mounted under the package tray that recirculated passenger compartment air over the inside of the rear window. Several custom lighting options were available: glovebox ($2.75), underhood ($2.95), trunk ($2.35), and back-up ($7.50). Cruise control—an option not generally associated with 1963 vehicles—was a dealer-installed option for both the Chevrolet and Chevy II at a cost of $87.00. It was a sophisticated device for the era. It controlled vehicle speed and would disconnect with a touch of the brake or clutch. Also available were litter containers, tissue dispensers, a compass, tool kit, hazard flashers, and several radio and antenna options.

A shrewd Chevy buyer in 1963 often asked for one of these dealer-installed extras to close the deal. After agreeing upon a price for his new Impala, he would say, "I'll take it—if you throw in a set of those deluxe full width floor mats." The buyer had to be careful not to ask for too much or the salesman would balk. Usually the salesman said, while scratching his head, "I don't know, lemme go talk to the sales manager." The salesman would grab the contract and walk to a back office beyond the view of the crafty buyer, and wait a few minutes. The salesman knew that giving up $10 for rubber mats was nothing compared to getting a $3,000 sale for that new Impala. The customer would leave with a big smile, driving his new Impala and gloating over how he had out-foxed the dealer on the mats.

Z-11 suspension was simple but functional—made for straight line passes only. All Z-11s came with a standard 4:11 positraction rear end. Wheels were 15x5 1/2-inch police models with 6.70x15 tires.

The body shell on the Z-11 started as a standard Impala two-door coupe (model number 1847). It carried all standard exterior and interior trim. The only exterior engine identification was the crossed-flag emblem on the front fender, bearing no tell-tale numbers above. The body was assembled without any of the sound deadeners usually applied to the inside of doors, hood, trunk, and underbody of a production passenger car. The sound deadener was deleted to help reduce vehicle weight.

A major part of the Z-11's diet plan was its aluminum body components. Both the front and rear bumpers were aluminum and the brackets attaching the bumpers to the frame were made of aluminum. Front fenders were also made of aluminum and were very frail due to the thin stock used. Most Z-11 fenders developed stress cracks in the area behind the front wheel at the rocker panel attachment due to the twisting of the body from engine torque. The hood and its inner reinforcement were also aluminum. These hoods were very flexible and few fit properly.

The Z-11 interior was standard Impala trim level with radio and heater delete and a tach redlined at 6,000. The shifter for the close-ratio transmission was a standard swizzle-stick design. Shipping weight for the Z-11 was 3,245 pounds, 145 pounds less than the standard Impala.

The use of the heavier Impala body begs the question, why not use the Biscayne two-door sedan body and save even more weight? The V-8 Biscayne was only 50 pounds lighter than the standard Impala. Chevrolet's goal behind this program was to draw customers into the dealer showroom. And if you were a salesman, what would you rather have them buy, a low-cost Biscayne or an Impala with lots of options?

Chevrolet's NASCAR effort with the 409s had limited success. They proved durable and ran well at the short tracks where the 409's quick-revving ability could be put to its best use. The 409's downfall was the lack of top speed on the super speedways. Chevy engineers came to the realization that the 409 was not as responsive to the usual performance

Blacking out the vertical grille bars was the most notable of exterior trim changes made to the 1963 Chevy IIs. In 1963, Chevy IIs were split into three series: 100, 300, and Nova. This is an example of the midrange 300 series four-door sedan, which stickered for $2,180 when equipped with a 120-horsepower, 194-cubic-inch engine. *GM Media Archives*

This 1963 Impala convertible is equipped with a 195-horsepower, 283-cubic-inch engine, as evidenced by the small V on the front fender. It had a sticker price of $3,024. Over 800,000 Impalas were sold in 1963, and 82,659 of them were convertibles. *GM Media Archives*

By design, Chevrolets have always had the look of a baby Cadillac, and the 1963 Chevy was no exception. Several styling cues from the 1961 Cadillac could be seen in the full-size 1963 Chevy. The full-length body character line that passed middoor made it look longer. The color-keyed lower body molding also accentuated the length. The tight egg-crate grille and general crispness of the lines echoed Cadillac. *GM Media Archives*

modifications as was the small-block. The Chevrolet engine group sent the message upstairs that it wanted to redesign the W motor or make a new one. From his days at Pontiac, Bunkie Knudsen knew the positive effect a successful racing program could have on sales, and in the summer of 1962, Knudsen gave the go-ahead to design the new engine.

In late July 1962, Dick Keinath headed up the small team of engineers that started work on what was to be a new 409-cubic-inch engine. Late in the design cycle, the displacement was increased to meet the new NASCAR maximum of 427 cubic inches. Keinath's team worked six and seven days a week around the clock on an engine no one outside of Chevy knew about. The engine was given the code name of Mark II. It was a design that took the best of the successful small-block and improved upon it. "We started with a clean sheet of paper," said Keinath. "But we had to save some stupid things like pan gaskets and some other seals."

One improvement over the small-block design was the elimination of Siamese exhaust valves (on cylinders two and four, and three and five). "The Siamese exhaust valves concentrated too much heat around those two valves," said Keinath. "Because of the excess heat, the valve seats would tilt, causing valve leakage and burning under extreme conditions—so I changed it."

Keinath is also responsible for the canted valve design of the Mark II. It was initially done to allow the use of larger valves. The ball-stud rocker arm was the key in allowing the valves to be tilted. Later, this valve placement in the head was nicknamed "porcupine" because of the way the rocker studs

protruded from the head in several directions.

During the winter months, Chevrolet was secretly running tests on the Mark II 427 at its Mesa, Arizona, proving grounds. Cars used for the tests were NASCAR clones complete with roll bars. Speeds in excess of 180 miles per hour were achieved and Chevrolet was poised to shake up the troops in February at Daytona.

The big shake-up occurred, however, in late January. General Motors officially announced it supported the AMA's ban on racing and no longer supported *any* kind of organized racing—officially or unofficially. Just prior to that announcement, the new 427s were delivered to five NASCAR competitors with a selection of spare parts. With the GM-imposed ban, the recipients of these engines didn't know if they had enough spare parts to last the season or if additional engines would be available.

In February 1963, five Impalas rolled into Daytona for the 500 with the new 427-cubic-inch engines under the hood. With the recent demise of racing programs, Chevrolet officials were mute about the new engines. Almost every conversation in the city of Daytona was about the new Chevy 427s. Their power

Anchoring Chevrolet's four-model 1963 line-up was the smartly restyled Impala Sport Coupe. It carried over the 1962 Impala's roofline, but had a completely restyled body. At the top of Chevrolet's 1963 offerings was the hot new Corvette Sting Ray. Chevy dealers loved to park one in the showroom to increase floor traffic. In between are the sporty Corvair Monza and the very practical Chevy II, two low-cost cars that offered plenty of value. All 1963 Chevy models had self-adjusting brakes as standard equipment, and all but the Corvair had a Delcotron alternator. *GM Media Archives*

output was far above that of their competitors. Rumors of their power were backed up by their track times. Junior Johnson was turning practice laps of 168 miles per hour. Top qualifiers were the 427 Chevys of Junior Johnson (163.68) and NASCAR rookie Johnny Rutherford (165.18), who sat on the pole. The 1963 Impalas were running 10 miles per hour quicker than the fastest Chevy in 1962. Competitors could only hope the new, unproven engines would not be durable. Two of the new engines did fail during practice, but all five were there for the start of the race.

When the green flag dropped for the 500, Junior Johnson was the rabbit everyone was chasing. Unfortunately, Johnson's hard running broke a pushrod and he retired having run only 26 laps. The race was dominated by the 1963-1/2 427-cubic-inch Fords and was won by Tiny Lund. Johnny Rutherford finished in ninth place, the highest-finishing Chevy.

The Z-11 and Mark II 427 died an early death due to General Motor's decision to abide by the AMA's 1957 decree to end involvement in all racing programs. Parts were scarce for both of these engines through the balance of the year. As automotive technology marched on, Chevrolet, without competitive engines, lost its biggest names in racing to the Ford and Mopar camps.

Performance cars were still available from Chevrolet, but couldn't compete at the top levels of racing. The factory efforts of both Ford and Chrysler were now in high gear. The engine that had stunned the competition at Daytona was not be seen again until the spring of 1965.

Chevy II

The 1963 Chevy II line-up remained unchanged from the 1962 models. Small technical changes were made to improve quality and durability. Like all the

Bill Mitchell liked fake vents and the new Corvette had quite a few. Nonfunctional twin depressions on the hood were fitted with aluminum grilles. Just to the rear of the front tire were two depressions—also nonfunctional. The knock-off aluminum wheels were a new option for 1963 and had some teething problems due to porous castings. *Corvette Fever*

other Chevy car lines for 1963, the Chevy II had self-adjusting brakes and a Delcotron alternator as standard equipment. Minor changes to exterior ornamentation were the only visible differences.

One bright spot for the Chevy II line in 1963 was the introduction of the Nova Super Sport option. The SS was available on both the Nova 400 Sport Coupe and convertible. This $150 package was similar to the Impala SS option featuring bucket seats, console, and optional floorshifted Powerglide. Fourteen-inch wheels were now optional on all models and required with the SS option, and the Nova SS wheel covers were the same tri-bar spinners as on the Impala SS. The SS option was popular—over 87,000 were sold. Had a factory-installed V-8 been an option on the 1963 Chevy II, sales would have been even better. That was remedied the following year.

Corvair

With much of Chevrolet's engineering talent working on the new Impalas and Corvettes, the Corvair, like the Chevy II, received little attention. The basic body was carried over from 1962 with only ornamentation changes. The only sheet metal change was the deletion of the two depressions below the back window on the coupe body. Corvair's full wheel covers were carried over from 1962, with the addition of black-painted accents.

Slow-selling Corvair station wagons were dropped from the 1963 line-up. Monzas were hot sellers, with the sporty coupe totaling almost half of all Corvair sales in 1963.

The 427-cubic-inch engine that powered the 1963 Z-11 Chevy was a stroked version of the 409. It featured new heads and a high-rise intake manifold. Covering the dual four-barrel carburetors was a large air cleaner that received its fresh air from a duct connected to the cowl. The cowl vent, at the base of the windshield, is a high-pressure area providing a mild supercharging effect at speed.

Chapter 5

1964 New "Mama Bear" Chevelle

Chevrolet and parent company General Motors had an outstanding financial year in 1964. Dividends paid on GM's common stock were $4.45 a share, an 11 percent increase over 1963. General Motors paid $1.2 million in dividends, the largest dividend any company had ever paid. Chevrolet sales for 1964 totaled 2.3 million units, another record and half of GM's 4.6 million in domestic production. In 1964 one out of every four vehicles sold in the United States bore a Chevrolet nameplate.

General Motors announced early in 1964 a two-year, $2 billion capital expenditure program. New plants were to be opened and existing factories and production equipment were to be modernized to keep pace with growing sales. General Motors wanted to be sure it had the ability to produce and deliver the cars the public wanted.

American cities were expanding horizontally, creating new suburbs and thousands of miles of new roads connecting work, home, and recreation. The auto-buying public was also growing, as the postwar

The new 1964 Chevelle was based on GM's A body, shared with the Pontiac Tempest, Olds F85, and Buick Skylark. It was the first Chevy with curved side glass, which offered stylists a great degree of design freedom. The small V on the front fender of this Super Sport indicates a 283-cubic-inch V-8. Late additions to the option list were the same 250-horsepower and 300-horsepower 327-cubic-inch V-8s as used in the Corvette and full-size Chevy. These engines were added to close the performance fissure created by Pontiac's hot-selling GTO option. *GM Media Archives*

57

The 1964 Corvette Sting Ray coupe carried the same sleek lines as the 1963. The Corvette's two medium-performance 327-cubic-inch engines remained at 250 and 300 horsepower. The two high-performance engines were now rated at 365 horsepower, with a single Holley four-barrel carburetor, and at 375 horsepower for the Rochester fuel-injected version on this coupe. *Corvette Fever*

baby boom had reached a car-buying age. They wanted more choices and cars that were fun to drive. Their parents were still buying cars and were now able to afford more upgrades and options. Consumers' demands for extras were high, and Chevrolet did its best to fulfill the needs of every car buyer.

Chevrolet was trying to permeate every niche in the American automotive market, and it was also building cars to counteract Ford's success in the market. Chevrolet didn't want Ford to have a sales advantage, leading to moves like one in 1962, when Chevrolet released the Chevy II to counter Ford's conventionally designed Falcon, which soundly outsold the Corvair. The Falcon was basic transportation, while the Corvair was seen as somewhat exotic. In 1962, when Ford released its Fairlane, Chevrolet had nothing with which to counter until the release of the 1964 Chevelle.

To stay in touch with the car-buying public, GM's marketing staff mailed two million questionnaires to citizens across the nation. They surveyed their likes and dislikes on current models and what changes the automakers should have made. This and other market research programs were used to

identify long-range trends in the automotive market. General Motors and Chevrolet continually evaluated the market to determine customer needs and tastes, an extremely difficult task.

Auto safety was a growing issue in 1964. Seatbelts, once optional, were now standard on all 1964 cars. Designers were also looking at the interiors of the new cars they were designing and making changes to improve passenger safety.

April 17, 1964, is a day the folks at GM will never forget. It was the day Ford Motor Company introduced the Mustang. From that day on, the automotive world was never the same. Even die-hard Chevy fans had a hard time disliking the Mustang. Ford's Lee Iacocca had taken the Falcon platform, hung some great-looking sheet metal on it, and turned it into an American icon. GM was caught napping, and for the next three years, pushed ahead at flank speed to produce a car with the same consumer appeal as the Mustang.

In September 1964, the UAW called a strike against GM after its collective-bargaining agreement had expired. The UAW leadership turned down the offer made by GM, which was similar to

Leading this pack of 1964 Chevys is the all-new Chevelle. Directly behind the Chevelle is a Corvair Monza coupe. The back row is led by a Corvette with a Chevrolet and Chevy II following. These five Chevrolet car lines accounted for record sales in 1964 of 2.3 million units. *GM Media Archives*

the contracts already signed with the other automakers. National issues were settled quickly, but local issues required weeks to settle. The earnings of GM's hourly workers were well above the average for U.S. manufacturing employees, according to the Bureau of Labor Statistics. In 1964, the average GM hourly worker earned weekly wages of $150.20—an increase of $5.86 over 1963. The strike lasted 45 days and idled 275,000 GM workers.

Genuine engineering accomplishments, while small, were seen across Chevrolet's car lines. In 1964, Chevrolet standardized its wiring color code for easier service. The Chevelle was the first Chevy to use the new, flat electrical harness, which was

molded in the shape of a ribbon instead of having a bundle of wires taped together. With this new shape it was much easier to route wires along a shallow depression in the floor pan.

In 1964 Chevrolet and General Motors realized that the automotive needs of America were growing. Customers wanted cars tailored to their requirements, cars in different sizes with a full range of options. Sporty cars sold well, and 25 percent of all cars sold in 1964 were two-door hardtops. Performance was on the minds of consumers too, as 69 percent of the cars were sold with V-8 power. In 1954, two out of three families owned a car. In 1964, that figure was up to four out of five. Also in 1954, families that owned two cars were only one in eight. In 10 short years that figure jumped to one in four. America wanted wheels and lots of them.

Chevelle

In 1964, Chevrolet introduced the Chevelle—Chevy's third new nameplate in four years. Chevrolet now had a "papa bear, mama bear, baby bear" car line-up, with the Chevelle sliding into the midsize mama bear position. It was based on the Fisher Body A body shared by the Pontiac Tempest, Olds F-85, and Buick Skylark. These models were the first GM passenger cars to have curved side glass, a feature that gave the stylists more freedom and creativity in their designs. Though based on the same platform, each division's offering was highly distinctive.

The Chevelle fit nicely in between the small Chevy II and the full-size Chevrolet. The Chevelle was an efficiently designed car, with its interior dimensions within an inch of those of the larger Chevy, while its exterior was a full 16 inches shorter. The Chevelle was often compared to the 1955 Chevy in size and appearance, and like the 1955, Chevelles were attractive, affordable, and good performers. Chevrolet enjoyed this comparison, as the 1955 Chevy was accepted by automotive enthusiasts as a classic, though not yet 10 years old.

Eleven Chevelle models were offered in three series: a base 300, an upgraded Malibu, and a Super Sport. The 300 was available as a two-door sedan or a four-door sedan, and each was offered with either a basic six or V-8 engine.

In addition to those models, the El Camino was again available after a three-year absence. When Chevrolet dropped the El Camino at the end of the 1960 model year, Ford continued its version, the Ranchero, in its Falcon series. Chevrolet did not want to miss any market segment, however, and the Chevelle El Camino was a perfect fit as a quasi-pickup. The popular El Camino was carried as a Chevelle model for the balance of the decade.

The styling of the 1964 Chevelle was looked upon by the critics as a classic. The large, open oval-shaped grille with its egg-crate pattern had the appearance of a scaled-down Impala. Sculptured quarter panels had just the right amount of bulge to lend a smoothness to the design. Front and rear wheel openings were of similar shape, with the rear cut lower for a more formal look. The oval shape of the front was carried over to the rear, and it was judged by some critics as a bit busy and not as clean as the front layout.

When the designers trimmed the SS Chevelle for 1964, they looked back to the simple elegance of the

The 1964 Chevelle was brought out to compete against Ford's midsize Fairlane. It was a full-frame car, 194 inches long with a 115-inch wheelbase. These dimensions were very close to the dimensions of the 1955 Chevy, and obvious comparisons were made. *GM Media Archives*

1962 Pontiac Grand Prix. Under the direction of then-Pontiac head Bunkie Knudsen, the chrome was stripped off, revealing the beauty of the Grand Prix's styling. Knudsen's magic was applied to the Super Sport Chevelle to create its distinctive look. Simple thin chrome strips outlined the fender, door, quarter, and wheel opening edges framing the Super Sport's sides. This outlining was exceptionally attractive against one of the many dark colors available. The Super Sport Chevelle was a far cry from five years earlier, when every GM vehicle was covered in chrome.

Powering the new Chevelle was Chevy's line-up of familiar six-cylinder and V-8 engines. The 194-cubic-inch 120-horsepower six, from the Chevy II, was the standard six. For an extra $40 the 155-horsepower, 230-cubic-inch six was the fuel-stingy upgrade. Transmissions for these engines were the basic three-speed manual, overdrive, and Powerglide.

Upon release, 1964 Chevelles were available with two versions of the 283-cubic-inch V-8. A 195-horsepower two-barrel carbureted version was the base offering, and a 220-horsepower four-barrel dual-exhaust version of the 283 was optional. Both V-8s were available with three-speed manual, four-speed manual, and Powerglide transmissions. The 195-horsepower engine could also be ordered with an overdrive transmission.

In April 1964, Chevrolet released the 250- and 300-horsepower versions of its 327-cubic-inch engines as options for the Chevelle. This was probably done in response to the overwhelming success of Pontiac's GTO. These engines were backed by the same transmission combinations as the 220-horsepower 283.

The rear axle design on the 1964 Chevelle was similar to those used on the 10 series Chevy trucks. It was much stronger than the standard passenger car rear axle and was identified by its 10-bolt cover. It was available with ratios matched to specific engines and transmissions; positraction was an option.

In 1964, Chevrolet made the Super Sport a distinctive model instead of merely an option to the Impala, as it had been in the past. In doing so, the Super Sport was upgraded considerably. The 1964 SS Impala now had its own interior trim that included unique door trim panels. The Super Sport bodyside molding was new and not a modified version of the standard Impala, as it had been in 1962 and 1963. The base price of this V-8-equipped Super Sport convertible was $3,196, a mere $161 increase over the Impala convertible. *GM Media Archives*

Four different frames of the same basic perimeter design were used for the 1964 Chevelle. One frame was used for the coupes and sedans, and one of equal length with two additional body mounts was used on the convertible. Frames under station wagons and El Caminos were 3.25 inches longer. Side rails were an open C section. Front and rear suspension mounting areas of the frame were fully boxed, and spanning the frame were three welded cross-members. The front cross-member was the largest and was fully boxed, and it supported the engine and front suspension. A large, deep channel section cross-member above the rear axle provided mounts for the coil spring pockets, shock absorber mounts, and upper con-trol arms. A smaller open section rail tied the rear of the frame together. One removable cross-member was used as a transmission support.

Chevelles sold well, as did all of the A-bodied cars GM offered in 1964. It seemed to be the right size and have a good mix of options to suit any taste. Performance enthusiasts were disappointed in the selection of engines available, but that was about to change. Bunkie Knudsen had been out-foxed by his old buddies at Pontiac with the GTO, but he got his revenge in 1965.

Chevrolet

After three years, Chevrolet management was finally convinced that the Super Sport was for real.

The most dramatic change to the 1964 Corvette coupe was the removal of the rear split window. This change was implemented prior to the introduction of the 1963 Coupes and was done to increase rear visibility. The coupe's left sail panel vent was made functional, bringing cool air into the compact passenger compartment. The Corvette's full wheel covers were restyled, maintaining a tri-bar spinner. *GM Media Archives*

In 1964, the SS became a distinct model instead of being an option to an Impala Sport Coupe or convertible. Until then, the Super Sport was rather limited in scope. In previous years, interior designers had wanted to do a unique trim for the SS but were restrained from doing so by cost. Only small changes and additions could be made. Interior badges, a console, and bucket seats could be added, but the rear seats were an Impala design. Changes were not allowed to the door trim panels either—too costly. Prior to 1964, exterior SS side moldings were identical to those of the Impala, with the exception of a swirl pattern insert replacing a painted insert. The 1964 Super Sports were trimmed with a unique full-length body molding.

Ford's sporty Galaxie 500 was selling well and Chevy kept the price low on the SS to steal sales away from Ford. The price differential between an Impala and a Super Sport was only $150—a small amount to pay for a very distinctive car.

Bunkie Knudsen's personal car in 1963 was a modified Impala Super Sport convertible. It was not unusual for Detroit executives to drive a production-based car with a special paint scheme or other custom touches that would be seen eventually in production cars. Knudsen's Super Sport had a one-

off interior trimmed in light turquoise leather. The design, created by the Chevrolet interior studio, carried the Y-pattern door scheme and pleated seats of the production 1964 Super Sport.

Exterior styling for the 1964 Chevrolet was more formal than the 1963 models. The grille opening was enlarged, fully surrounding the headlights in a large oval shape. Small circular taillights in groups of either two or three, depending on the model, graced the rear. Doors and uppers were carried over from 1963 with attractive side moldings masking the carryover doors in the Impala and SS models. To keep costs in line, little chrome-plating was used. Most of the exterior trim was anodized aluminum.

Chassis design for the 1964 Chevy was the same basic layout that supported the 1958 Chevy. Small improvements were made from year to year, but now Chevy's chassis was dated.

Engines and transmissions for 1964 were the same as 1963. The base engine was the 235-cubic-inch six-cylinder. The tried-and-true 283 was at the bottom of the V-8 list of engines with two optional 327s available. The 340-horsepower 409 was received well in 1963 and was again an option. In an era of inexpensive fuel, the 340-horsepower 409 was an easy item to select when ordering a fully

The 1964 Chevy IIs carried over the 1963 body, but added new exterior identification, new interior trim, and a refined egg-crate design grille. This deluxe Nova model is distinguished by the bright molding that extends over the rear wheel openings to the rear bumper. A V-8 was finally available as an option in the Chevy II in 1964, but it was only a 283 with a two-barrel carburetor. *GM Media Archives*

The 1964 Chevy was an updated version of the 1963. Doors and body uppers were carried over with new front and rear sheet metal added. Engine and chassis configuration remained the same. *GM Media Archives*

optioned Chevy. The two high-performance 409s were identical to those offered the previous year. Unfortunately, they lacked the power of Chrysler's Hemis or Ford's high-risers, so anyone who wanted to be competitive in auto racing was not running a Chevy in 1964. A few die-hards tried, but it would be years before Chevy was in the winner's circle again.

Chevy II

The Nova for 1964 remained essentially unchanged from the 1963 model. The only news of note was the addition to the option list of the 283-cubic-inch V-8 and a four-speed manual. Unfortunately, the only 283 offered was the two-barrel version rated at 195 horsepower. With the good news of the V-8 in mind, Chevrolet dropped a bomb by eliminating the sport coupe and convertible from production along with the Super Sport option. By midyear the sport coupe and the Super Sport option were reinstated, but the convertible was not. Chevrolet no doubt felt that the upgraded Nova would hurt Chevelle sales.

Corvair

The appearance of the 1964 Corvair was very similar to that of the 1963. This was the fourth and last year for this body, first introduced in 1960. Chevrolet and all of the Detroit automakers had come to realize how expensive yearly sheet metal changes were, especially on the smaller car lines where profit margins were thin. New moldings, nameplates, and exhaust grille were used to distinguish the 1964 Corvair from its predecessor.

Corvair's swing-axle rear suspension was carried over into 1964 with one important change: A single transverse leaf spring was added. Its mounting was similar to the one on the Corvette, and this addition reduced the effect of rear-wheel tuck-under on hard cornering. The balance of the rear suspension stayed the same except for the reduction in spring rates of the coil springs to compensate for the added load-carrying ability of the leaf spring. Finned brake drums were added as standard equipment on all 1964 Corvairs.

All that was new for the 1964 Corvair was under the hood. Three horsepower levels (95, 110, and 150) were available out of the increased-displacement flat six. Cubic inches were increased to 164 with the addition of .34 inch to the stroke of the forged crankshaft, while the bore remained the same. A new, higher-lift camshaft was added along with new cylinder heads. All engines now used the I-beam section connecting rods previously used only on the turbocharged engines. To compensate for the increased displacement, all Corvair carburetors were recalibrated.

Corvette

Sophomore models are rarely thought of as being all that special. Not so with the 1964 Corvette. Chevrolet made several changes in the Corvette, all of which improved the car significantly in both performance and comfort.

Interior ventilation on the 1963 coupes had been less than adequate. Driving a 1963 coupe with an automatic transmission without air conditioning on a warm, humid day was torture. The heat radiating from the transmission and the lack of adequate air circulation brought sauna-like conditions to the Corvette's confined interior. Chevrolet's solution was to add a blower in the left side of the rear luggage

Chevrolet released this photo of the Corvette assembly line in 1964. Chevrolet's photo lab retouched a 1963 photo by removing the grilles from the hood depressions to make the Coupe in the foreground look like a 1964. What the photo reveals is that the St. Louis Corvette assembly plant was not nearly as up-to-date as the cars it produced. *GM Media Archives*

The 1964 Malibu Super Sport was a distinct model like its big brother, the Impala Super Sport. It featured thin bodyside moldings that outlined the side of the vehicle. These thin moldings added a clean look of elegance to the car. The Malibu Super Sport's tri-bar wheel covers were the same as the ones on the Impala SS. *GM Media Archives*

There were no technical advances of note on the 1964 Chevy, except for the addition of seatbelts as standard equipment. The chassis was a dated X-frame design first seen on the 1958 Chevy. The overall styling was boxy compared to that of the 1964 Fords, but the Chevys sold well due to the value customers got for their money. This Bel Air four-door sedan could easily seat six and offered 19 cubic feet of trunk space. *GM Media Archives*

compartment. This blower drew air from the rear of the luggage compartment and pushed it through a series of ducts, exhausting it through the now functional gill vents on the left sail panel. The rear vent was activated by pulling the rear vent control knob on the instrument panel. The first detent opened the duct, and subsequent detents activated three speeds of the blower. Additional passenger compartment insulation was added to reduce heat and noise.

Body-to-frame mounts were revised for both the coupe and convertible. Coupes had their body mount locations reduced to six from the previous eight, and all six mounts used rubber biscuits for isolation. Convertibles retained the original eight mounts, but only the four midship mounts were converted to the rubber biscuits.

The split in the Corvette coupe's rear window was now history, and a new one-piece rear window was in place. As much as the automotive journalists of the day would like to have taken credit for pressuring Chevrolet to remove the split, the one-piece glass had already been scheduled for 1964 models prior to the release of the 1963 coupes. Chevrolet eliminated the rubber weather strip that held the previous back windows and installed the new rear window with a caulking compound that glued it in place.

Other exterior changes of note included the removal of the fake grilles in the hood depressions and the addition of new, full wheel covers. Problems with porous castings that plagued the 1963 Corvette

The Chevelle Malibu was offered in either a four-door sedan (as shown), a two-door sport coupe, a convertible, or in two station wagon models. Like all 1964 Chevelles, this four-door was available with a long list of options. *GM Media Archives*

knock-off aluminum wheels were solved for 1964. This attractive option listed for $322.80 and offered the purchaser five 15x6 true knock-off wheels.

Under the hood, changes were made to increase the output of the two high-performance 327 engines. New larger valves and a longer-duration camshaft were added to the two solid-lifter engines. This increase in breathing upped the fuel-injected version to 375 horsepower. The high-performance carbureted version also gained a new intake manifold and Holley carburetor, increasing its output to 365 horsepower. The two hydraulic-lifter 327 engines remained the same, at 250 and 300 horsepower.

Chapter 6

1965 Sleek Style and a New V-8

Demand for Chevys in 1965 was relentless. Even the UAW strikes in late 1964 didn't hurt new car sales. Old stock sold out and demand for new cars began to build. And in 1965, Chevrolet had a lot to offer. Both the Corvair and full-size Chevy were completely redesigned. The design of both models was cleaner and more integrated than those of any passenger car Chevrolet had ever produced. Midyear releases of a new engine and a new upscale Chevrolet model continued to draw showroom traffic. Technical improvements and minor face-lifts covered the balance of the 1965 Chevrolet line-up.

Sales records were set again in 1965 with 2.3 million cars sold—almost half of those bore the Impala designation. Hot on Chevy's heels was Ford, with sales of 2.1 million units, a 36 percent increase over its 1964 sales, due in a large part to the sky-rocketing success of the sporty new Mustang.

Chevrolet batted nearly 1,000 in 1965 by introducing a beautifully styled new body on the full-size Chevy, the classy Caprice option, a new high-performance engine, a new automatic transmission, and a new SS396 Chevelle. Sales were spectacular and life was good in Detroit. Chevrolet was still the low-price leader and defied the competition to outdo them. Chevy performance enthusiasts were ecstatic over the release of the 396-cubic-inch engine. The

The stunning appearance of the 1965 Chevy was a bold departure from previous Chevrolet designs. To add to its beautifully sculpted body, Chevy offered 13 new color selections. Evening Orchid, Crocus Yellow, and Glacier Gray were exclusive to the Impala and the Impala Super Sport. *GM Media Archives*

Corvette interiors received a lot of attention for the 1965 model year. The seats were redesigned and leather coverings were offered in any of seven interior colors available (silver is shown). The standard Corvette steering wheel was a simulated woodgrain (shown here); the optional wheel was teakwood. The carpet was a one-piece molded design and the console was a new, brightly finished diecasting. Corvette door panels were now a molded design (vinyl backed by foam), a first for GM. These new panels gave the stylists freedom to create smoother sculpted designs with integral armrests.

The 1965 Corvette Sting Ray body was the same as the 1964, with two exceptions. Three functional vertical louvers in each front fender were added, and the hood depressions were deleted.

two-year high-performance drought had not hurt sales, but Chevy's performance image had suffered. Also flagging were the sales of the Chevy II and the newly restyled Corvair. Nevertheless, in 1965 Chevrolet became the first car company in history to produce three million cars in one calendar year.

Chevrolet

A clean sheet of paper was rolled out for the design of the 1965 Chevy. A completely new body with dramatic styling rode on a full perimeter frame. Curved side glass, first seen on the 1964 Chevelle, gave stylists additional freedom to create Chevy's beautiful new shape.

Design criteria for the new Chevy called for a longer, lower car. Trunk space was to be maintained with a deep well. Jim Premo, then Chevrolet's chief engineer, determined that the trunk well should be deep enough to hold a dairy farmer's standing milk can. Another area of concern for engineers was the pancake hood. It was bounded on four sides by fenders, a nosepiece, and cowl vent, creating an assembly and alignment nightmare. The location of the new blade bumper, which was out of alignment with the frame rails, created another challenge for Chevrolet's engineers.

The original exterior styling of the 1965 was very conservative and only a short step away from the design of the 1964 Chevy. The new upper body had a fastback design, but the body had an unimaginative extruded side. Bill Mitchell and Bunkie Knudsen approved the design and work began. Irv Rybicki, Chevrolet's chief designer, was not happy, however,

with the stodgy look. He directed a second, unautho-rized design without the knowledge of Mitchell or Knudsen. The new design ran concurrent with the original, using the same upper body, but featured a dramatic new lower body, sketched by Emil Zowada. His renderings were based on a prominent rear fender form. The last portion to be resolved was the front end, which appeared too static when viewed from the side. Designer Henry Haga resolved the problem by putting a forward lean into the design. The 1965 Chevy front end broke away from the traditional layer-cake look of hood, grille, and bumper. Below the nosepiece, the grille was split by the slender blade bumper. Irv Rybicki's team had created a new design for the 1965 Chevy that was far superior to the original.

Once completed, Rybicki revealed the alterna-tive design to Mitchell and Knudsen. Knudsen said there was no alternative—*this* was the design. Mitchell was also pleased with the design, but not entirely ecstatic at being surprised. In a private meeting with Rybicki, he warned him never to pull a stunt like that again. Mitchell hated surprises.

The remaining problem with the new design was that it had not been approved by GM's Board of Directors, which had already approved the earlier design. The board was a conservative group, reluc-tant to make changes, but Knudsen built a strong case in his presentation and the board approved the change. Such a major revision so late into the design cycle had never been approved before in the history of GM, yet this decision was a first, over which there were ultimately no regrets.

Supporting the Chevrolet's beautiful new body was a full perimeter frame. An all new coil-spring front and rear suspension was added. Similar to the Corvair's design, the front suspension featured a sin-gle bushing inner mount on the lower control arm, supported by a horizontal strut rod. The upper arm was a conventional A-frame design. The new Salis-bury type rear axle was borrowed from the Chevelle. It was mounted with two lower control arms and one upper on the right side. Chevrolets equipped with any of the optional V-8s were given an additional upper control arm on the left side. Brakes were carry-

The limited-edition Z-16 Chevelle was Chevy's answer to Pontiac's GTO. Only 201 Z-16 Chevelles were built in 1965. They were all highly optioned Super Sport models with a 375-horsepower 396-cubic-inch engine. The bulk of the Z-16s went to major automotive magazines for road tests, to GM executives, or to high-profile personalities like Dan Blocker (original owner of this Z-16), a star of television's *Bonanza* series (sponsored by Chevrolet). *Tom Shaw/ Muscle Car Review*

over Bendix-type self-adjusters, while braking options were power-assist and metallic linings.

At new car introduction, Chevy's 1965 power-train line-up was the same as the 1964 line-up. The only exception to that list was the now-absent 425-horsepower dual-quad 409. By midyear the engine line-up changed dramatically, as both remaining 409s were dropped and the new 396-cubic-inch engines were added. This new engine release bred as much excitement as did Chevrolet's first V-8 10 years earlier. Dick Keinath's 1963 Mystery 427 design was brought back from the dead. This time it was downsized by 31 cubic inches to comply with GM corporate standards—400-cubic-inch maximum displacement. Keinath originally designed the engine with a 427-cubic-inch displacement that could easily be upped to 454. Reducing the cubic-inch displacement to 396 was a simple process of reducing the bore from 4.25 to 4.094.

Two different 396s were offered in the Chevrolet passenger car, a 325-horsepower street version and a 425-horsepower powerhouse. The two engines were similar in exterior appearance, had the same bore and stroke (4.094x3.76), but inside were very different. The 425-horsepower version had a forged crank and pistons, while the same components in the 325-horsepower engine were cast. Four-bolt mains were used in the 425, and there were two-bolt mains in the 325. A long-duration solid-lifter cam was used in the 425 and a much milder hydraulic-lifter cam was in the 325. Exhaust valves on both engines were of 1.72-inch diameter with intakes of 2.06 inches in diameter on the 325 and 2.19 inches in diameter on the 425. A mixture of both Holley and the new Rochester spread-bore four-barrel carburetors appeared on the 325 engines. The 425 was only available with a large Holley four-barrel on an aluminum intake manifold. All 425-horsepower engines were

The full-size Chevy was completely redesigned for 1965 and featured a prominent rear fender profile. The new Chevy sat on a full perimeter frame with new suspension, rear axle, and steering. The Impala two-door hardtops were the most striking due to their fastback roofline. This Regal Red Impala Super Sport is powered by one of two 396-cubic-inch engines released midyear. Available as an option with the 325-horsepower 396 was the new Turbo-Hydra-matic 400 transmission. *Don Schwarz*

equipped with transistorized ignition systems, a magnetic pulse system that was available at extra cost on all optional V-8s.

Also introduced in mid-1965 was Chevy's second automatic transmission, the three-speed Turbo-Hydra-Matic 400. The Turbo 400 was only available in the full-size passenger car with the 325-horsepower 396. Its design was based on the hydromatic transmissions in Pontiacs and Oldsmobiles, and it was a perfect match for the powerful new 396 engine.

Knudsen had a habit of wandering through the studios early in the morning before the designers arrived. While nosing around one morning, he found sketches of a design based on the Impala. It was a luxury interior design done in response to Ford's upscale LTD. Knudsen liked the sketches and directed interior designer Steve McDaniel to design upscale trim and install it on a 1964 Impala four-door hardtop. This smartly trimmed black Impala became the prototype for the midyear 1965 release of the Caprice.

The rear of the 1965 Impala featured two sets of three taillights, a Chevrolet styling cue first seen in 1958. The two lights on the left are set onto the sloping surface of the rear of the deck lid, and the one on the right is mounted in the rear fender's end cap. *Don Schwarz*

OK Used Cars

Chevrolet was selling a lot of new cars in the 1960s, and as a result, Chevrolet dealers acquired as trade-ins a large number of quality used cars, usually Chevys. Used Chevrolet vehicles held their value better than most of the competition. In 1965, the average used 1962 V-8 Impala sold for $1,625, about 57 percent of the original retail price. By comparison, the average 1962 Ford Galaxie was returning about 51 percent of its original price, a 1962 Dodge only 46 percent.

In an effort to increase consumer confidence in its used cars and to move them off the lots, Chevrolet and its dealers offered a warranty on their used cars. The plan was called the Chevrolet OK Used Car Warranty, and under this arrangement, the dealer selling the car accepted responsibility for 50 percent of the cost of parts and labor for any repairs needed during the first 30 days. For the next two years, the dealer picked up 15 percent of the cost of repairs. Under the OK warranty, repairs had to be made by the issuing dealer's repair facility and at standard retail price.

Participating Chevrolet dealers pledged that all OK Used Cars would be honestly represented, fairly priced, and in good condition. The dealer also gave the buyer a written promise that each OK car had been inspected, road tested, and that necessary repairs had been made prior to the sale. Excluded from the OK warranty were tires, tubes, glass, radio, air conditioners, and any damage due to an accident or misuse. Tires were made available at a 25 percent discount during the first 30-day period only.

Ford and American Motors also offered used car warranty plans. All tried to build customer confidence and showroom traffic. In the 1960s, used car dealers were doing their best to establish credibility to a trade historically maligned.

The Caprice option to the Impala was a midyear release. It featured upgraded interior appointments, a vinyl roof, extra sound insulation, and special ornamentation. The Caprice was introduced in response to Ford's upscale LTD model. The name *Caprice* was appropriated from a posh New York restaurant frequented by Bob Lund, then Chevrolet's general sales manager. *GM Media Archives*

Naming a car model is not a simple task. Bob Lund, Chevrolet's general sales manager, offered the name "Caprice" to Bunkie Knudsen as a possible name for the new, upscale Chevy. Caprice was the name of a classy restaurant Lund frequented while in New York. The design staff hated the name, feeling it sounded too feminine. A list of names for the new model was generated, but Caprice stayed at the top. During a design review with Knudsen and chief Chevrolet interior designer Don Schwarz, Irv Rybicki tried one last time for a different name. Knudsen responded sharply, "I don't care if you call it a Schwarz; I'll sell it!" Knudsen went on to explain that when he took over Pontiac in 1956, its customer base was made up of "librarians and spinsters." He took the chrome stripes and Indian head off the Pontiac hood and went racing. "We changed the image; we didn't change the name," Knudsen said. The name Caprice stayed, and as Bunkie Knudsen predicted, it sold well.

Impala Super Sport production reached an all-time peak in 1965 of 243,114 units. Bucket seats, a floor-mounted shifter, and unique tri-bar full wheel covers distinguished it from the standard Impala's trim.

Corvette

As in 1964, only minor changes were made to the exterior of the Corvette. Most noticeable was the removal of the two horizontal depressions from the front fender in the area just behind the wheel. They were replaced with three functional vertical louvers, allowing hot air to escape from the engine compartment. The twin hood depressions were removed and a new grille was added. Other small changes in ornamentation rounded out the exterior.

GM engineer Alex Mair, who in the early 1960s was in charge of truck engineering, visited an auto show in England and noticed that quite a few of the cars and trucks had disc brakes. He did further investigation and determined the reason was that the mountainous terrain of Europe demanded better braking systems. When he returned to Detroit, he prepared a report on disc brakes for all the engineering groups within GM. The only person to take advantage of Mair's report was Duntov, who promptly commenced work to put them on the production Corvette.

In 1965, disc brakes were fitted to all four wheels of the Corvette. The caliper was a split assembly with two pistons on each side. Each pair of pistons acted on a brake shoe, more commonly called a pad. When activated, these pads squeezed a ventilated rotor.

Two different master cylinders were used on the 1965 Corvette. Cars without power brakes had a single-piston Bendix-type with a large reservoir. Power brake-optioned Corvettes had a Delco Moraine split master cylinder. The front reservoir and piston supplied the front brakes, and the rear reservoir and piston supplied the rear brakes.

Service of the new disc brake pads was simple. Each pad was grooved to gauge wear. When the groove disappeared—it was time for new pads. Changing brake pads on a Corvette was simple. Once the wheel was unbolted and removed, all the technician had to do was remove the cotter pin that retained the pad's guide pin, slide out the pin, and pull out the two worn pads. Once the four pistons were retracted back into the caliper, the new pads slid in place. The new brakes were so well designed, it took more time to jack the car up and remove the wheel than it did to replace the pads.

The parking brake was housed within the hub of the rear rotor. It featured two small brake shoes

The Corvair was dramatically restyled for 1965. All four-door models were pillarless hardtops. A new, solid rear suspension, similar to the Corvette, replaced the old swingarm design. Chevrolet executives expected the Corvair to give the Mustang a run for its money until safety issues sucked the oxygen from the atmosphere surrounding Chevy's sporty little car. *GM Media Archives*

very similar to a standard self-energizing brake. When the handle for the parking brake was actuated, it pulled a cable, expanding these miniature shoes against the inner surface of the rear rotor's hub. Adjustment and replacement of these shoes was not as simple as changing the pads.

With the new disc brakes, the Corvette finally had a braking system that was equal to the rest of the car. It was because of Duntov's installation on the Corvette that General Motors gained the experience needed to eventually offer disc brakes on all other GM car lines.

Besides braking power, 1965 Corvettes also got a big dose of horsepower. The base engine was the 250-horsepower 327. Optional was the 300-horsepower 327, and like the 250, it was a virtual carryover from 1964. New on the Corvette's option list was a 350-horsepower 327. It was identical to the 365-horsepower engine in every way except it used a hydraulic-lifter camshaft instead of a solid-lifter cam. The two higher-rated 327s, the Holley carbureted 365-horsepower and the fuel injected 375-horsepower version, were also carryovers from 1964.

Rumors of the Chevy II's demise were widespread in 1965 since Chevy's economy class sedan never sold up to expectations. Chevy II models ranged from economy four-cylinder sedans to this Nova Super Sport coupe, which could be ordered with a 300-horsepower 327-cubic-inch engine and a four-speed transmission. *GM Media Archives*

Standard equipment on all 1965 Corvettes were four-wheel disc brakes. These brakes were designed under the direction of Zora Arkus-Duntov to bring the Corvette's braking performance up to the levels of its engines and chassis. *GM Media Archives*

Available as a midyear release for the Corvette was the same 425-horsepower 396-cubic-inch engine that was available in the full-size Chevy. The Corvette version was identical to the passenger car version with the exception of a larger oil pan.

To satisfy the need for additional hood clearance dictated by the taller 396 engine, Chevrolet designers added what Chevrolet stylist Larry Shinoda called "the python that ate the pig." A squarish dome was added to the center of the Corvette's hood. Chrome louvered vents on each side of the bulge allowed a little extra heat to escape from under the hood. The new hood design made a strong visual statement about the powerful engine underneath.

The 396 has been unfairly accused of being the engine that killed Corvette's fuel-injection option. When it released it in March 1965, Chevrolet said the 425-horsepower 396 was to replace the Holley carbureted, 365-horsepower 327. But both the 375 and the 365-horsepower 327 engines were available through the end of 1965 production.

Since its introduction in 1957, Corvette's Rochester fuel injection had been given an undeserved bad rap as being difficult to tune and service. Sales of FI-equipped Corvettes had been steadily dropping over the years, and only 6 percent of all Corvettes sold in 1965 were equipped with fuel injection. That equates to less than 1,500 cars, a small number compared to Chevrolet's overall 1965 sales of 2.3 million cars. And with the availability of the new big-block engine producing 50 more horsepower for almost half the cost of the FI, it was

Chevrolet took full advantage of the legendary status of its small-block engine when it released its new 396. The 396 was offered in three horsepower ratings in three different car lines. In the full-size Chevrolet, 325- or 425-horsepower versions could be ordered. A 375-horsepower version was available in limited quantities in the Z-16 Chevelle. The Corvette offered only the 425-horsepower powerhouse. The Mark IV 396 replaced the 409s, which had reached the limit of their horsepower potential.

Nineteen sixty-five would be the last year for Corvette's fuel injection option. It was an expensive option and its sales had been in steady decline for years. The midyear release of the new Mark IV 396 rated at 425 horsepower, which cost half as much as the FI option, would seal the fate of the Rochester unit. This Corvette's Nassau Blue color was one of eight Magic-Mirror exterior finishes available for the Corvette. *GM Media Archives*

inevitable that Corvette's fuel injection was about to become history.

In addition to the new engines released in 1965, Chevy added another option that helped to express the masculinity of the Corvette: side-mounted exhaust, more commonly called side pipes. For a reasonable cost of $134.50, the owner of a 1965 Corvette could order side pipes with any engine or transmission combination. Side pipes were simply a pair of straight-through chambered pipes that ran from the engine's exhaust manifolds along the rocker and exited ahead of the rear wheel. The pipes were shielded with an anodized aluminum cover to protect passengers during entry or exit. Side pipes created an unmistakably loud, throaty resonance. A Corvette with side pipes driven for an extended period of time left the driver with a distinct ringing in the ears from the high-decibel level.

Chevrolet designers had been given the go-ahead to redesign the 1965 Corvette's passenger compartment. For the 1965 model year, interior designers were given the freedom to use a chrome-plated diecasting for the console plate instead of the cheaper anodized aluminum used on the 1963 and 1964s. For quite some time the designers had

wanted molded door trim panels. This type of panel had a skin of vinyl, backed by foam, which was molded in a die. Molded panels gave Chevy stylists more freedom in their designs. Sculpted designs and armrests could be molded into the panel instead of added on like an afterthought. Fisher Body objected to molded panels primarily on the basis of cost, but designers found a loophole in the system: Fisher Body did not have responsibility for the Corvette or trucks and couldn't officially say no. The designers forged ahead with molded door panels and added molded seatbacks, too. Chevrolet's interior designers used the Corvette's molded door panels as a proving ground for new ideas on a low-production vehicle. Once proven, these innovations filtered out to the rest of the car lines.

New and brighter paint hues were starting to appear, using the Corvette as the color test palate. Medium Blue was a particularly striking and bright color released in 1965. It was a color that had first appeared on a car driven by Bill Mitchell a few years earlier. According to Don Schwarz, head of Chevrolet's interior and color studio in 1965, "Mitchell had simple taste in colors: red, white, blue, and silver. To Mitchell, no other colors existed."

Protect-O-Plate

With the start of the 1965 model year, Chevrolet introduced its Protect-O-Plate system. It was developed to assist the dealer in faster and easier preparation of warranty forms. Additionally, the plates were used for dealer repair orders and customer follow-up information cards.

The Protect-O-Plate was an aluminum plate the size of a credit card that was glued to the back of the owners manual. Protect-O-Plate codes listed all options and the vehicle serial number for each car. An area was left blank for the dealer to attach an embossed tape with the owner's name, address, and date of sale. When warranty work was performed, the Protect-O-Plate imprint was made onto the warranty form with one fast swipe, leaving behind accurate owner information.

Today, Protect-O-Plates are a valuable tool for the restorer who wants to determine the exact options with which an older Chevy was originally equipped. Sadly, most of the plates have been discarded, leaving only the body data plate as a source of information.

The 1965 Chevelle received new front sheet metal that featured a slightly more rounded look than the 1964 models. This Malibu Sport Coupe was one of 12 Chevelle models available in 1965. *GM Media Archives*

It was an exciting year for Corvette enthusiasts: lots of new optional engines, disc brakes, side pipes, and a redesigned interior, and the Corvette again set sales records, selling 15,376 convertibles and 8,186 coupes.

Chevelle

The 1965 Chevelle received a minor face-lift that rounded and softened the lines of the successful 1964 Chevelle body. A new hood, grille, and front fenders were added; larger taillights were installed; and the back-up lights were relocated to the rear bumper.

The Chevelle's performance image had its doors blown off in 1964 when the extremely successful GTO hit the market. In an attempt to save face, Chevy installed the 250- and 300-horsepower 327s in the 1964 Chevelle as a midyear addition. These engines helped, but they were still short of the GTO's 325-horsepower base engine.

Chevrolet's best effort to counter the GTO in 1965 was the addition of the 350-horsepower 327. This engine was mechanically identical to the Corvette engine and was available on all models, standard with a three-speed manual transmission or optional four-speed manual. This engine was a big improvement, but it lacked the GTO's killing power on the street or in the showroom.

Chevrolet's knockout punch to the GTO was delivered on February 15, 1965, with the limited release of the Z-16 Chevelle. The Z-16 was a fully optioned Chevelle Super Sport with a new high-performance 396 engine. Chevrolet took the 425-horsepower engine, replaced the cam with a slightly tamer hydraulic-lifter version, dropping the horsepower to 375, and backed it up with a four-speed transmission.

All of the Z-16s built in 1965 were Super Sport hardtops. A long list of options was fitted to each of the cars, most notably heavy-duty suspension, tilt

In previous years, car roofs had been set on top of the body with little design integration. With the help of curved side glass, Chevy designers created a new highly integrated fastback roofline on the 1965 Chevy two-door Sport Coupes that dramatically changed that trend. The absence of side moldings on the Impalas focused attention on the overall beauty of the sheet metal. *GM Media Archives*

wheel, stereo radio, Firestone gold-line tires, and simulated mag wheel hubcaps.

Only 201 of these special SS models were built. In a public relations move similar to the one orchestrated for the 1961 409 SS Impala, they were immediately given to major automotive enthusiast magazines for road tests. A few Z-16s were driven by high-profile television personalities, VIPs, and Chevrolet executives.

Bunkie Knudsen had done it again with the Z-16 Chevelle. He built a car that created excitement and gave it to people he knew would be excited about it. Not one piece of sales literature was ever printed for dealers and no ads were ever placed in enthusiast magazines. Nevertheless, there was a big demand for Chevelles, especially those with a 396 engine.

Chevelle sales were at record levels in 1965, soundly outselling its major competition, the Ford Fairlane. Almost half of Chevelle sales were of the sporty Malibu and Super Sport models.

Chevy II

The 1965 Chevy II was available in three series: the 100, the Nova, and the Nova Super Sport. All three series had the same sheet metal first offered in

1962. A new grille and headlight trim were featured on the front end, and new, larger taillights and exterior ornamentation was added to the rear.

Six engines were available in the 1965 Chevy II, the most potent being the 250- and 300-horsepower 327s. These two robust V-8s, along with the steadfast 283-cubic-inch V-8, were all available with a standard three-speed manual transmission, optional Powerglide, or optional four-speed manual. Chevy IIs, with their light body and good selection of V-8 power, should have been big sellers. They weren't. Instead, Chevy IIs languished on dealer lots while customers snapped up Impalas and Chevelles. Selling only 122,800 units, barely half of what Ford's Falcon sold in 1965, Chevy IIs had the distinction of being the only Chevy model in 1965 with declining sales. Rumors persisted that the efficient little car would soon be dropped from Chevy's lineup. At a press conference during the 1965 Chicago Auto Show, Bunkie Knudsen stated, "We have no intention whatsoever of dropping the Chevy II."

Corvair

In 1965, the Corvair received a stunning new body. It also received bigger brakes, an improved rear

suspension, and more horsepower. Chevrolet was trying to equal Ford's Mustang by redesigning the 1965 Corvair.

Exterior design work for the 1965 Corvair started in late 1962 in a studio known as Advance Three. Ron Hill and Paul Gillen headed the design team for the second-generation Corvair, and they borrowed the concave rear treatment from the 1962 Monza GT show car. The rear window design was very similar to one done on a Pininfarina Corvair in 1960. A horizontal character line ran from the front edge of the hood, down along the side, washing out at the end of the quarter panel. This line made the car appear longer and lower. The style was definitely Italian. The 1965 Corvair's voluptuous curves could have easily come out of the Ghia or Pininfarina studios. Overall it was very well-proportioned, a difficult feat for a

rear-engine car. The Corvair, like the 1965 Impala, had been stripped of excess chrome and used the contours of the sheet metal to assert its beauty.

All Corvair models in 1965 were two- and four-door hardtops with curved side glass and no door pillars. Corvair model identification was divided among three lines: 500, Monza, and Corsa (formerly called the Spyder). The Corvair's wheelbase remained unchanged at 108 inches, but the overall length grew by 3 inches. It was a fraction of an inch lower and 2 inches wider than the previous model.

Handling was the other big Corvair news. Minor changes were made to the front suspension in the areas of attachments and adjustments. These changes altered the geometry slightly to augment the transformation in the rear. The entire rear suspension was redesigned and featured a design similar to

Little Bo Peep appears to have found her sheep as well as an inviting new red Monza convertible. The concave rear theme of the new Corvair was taken from the 1962 Monza show car. All Corvairs featured four taillights, reserving the inboard position for back-up lights. *GM Media Archives*

The Corsa was the top Corvair model offered in 1965 and was only available in a convertible or Sport Coupe with the 140-horsepower 164-cubic-inch flat-six as the standard engine. The turbocharged 180-horsepower engine was optional only on the Corsa. *GM Media Archives*

the rear suspension on the Corvette, with half-shafts, trailing torque arms, and strut rods. Rear coil springs were retained. These changes improved handling immensely, and the overall ride improved due to the selection of new spring rates. Brakes were upgraded using larger drums from the Chevelle. The Corvair now handled well and looked great.

Unfortunately, the damage to the Corvair's reputation had been done in earlier years and now Chevrolet was starting to pay. A lawsuit had been filed against General Motors by a woman who lost an arm in a Corvair accident. The case was settled out of court for $70,000 by the insurance firm representing General Motors. GM's lawyers were upset about the settlement, especially when the plaintiff's lawyers gloated over the victory, stating it was the first of 30 such cases they had against GM.

In 1964, a young lawyer named Ralph Nader was hired by a U.S. senator to write a brief on what the federal government should be doing in the area of auto safety. Nader worked for a year on his project and used articles from enthusiast magazines and Society of Automotive Engineers (SAE) reports to formulate his conclusions. A portion of his research was used as a basis for a scathing article on the Corvair for the November 1964 issue of *The Nation*. This article became chapter one of his book, *Unsafe at Any Speed*. Until Nader's article and book were published, GM's legal battles over the Corvair were merely regional news. Now it was fodder for the national news media and more deleterious attention was being drawn to the Corvair.

Chapter 7

1966 **Safety First**

The economy, while still growing, was in some turmoil in 1966. The war in Vietnam was going full tilt and social unrest was beginning to mount. These instabilities affected overall car sales, reducing industry volume to 8.8 million units. The hard pill for Chevrolet to swallow in 1966 was being outsold by Ford. The margin was a slim 5,000 units, but it gave Ford bragging rights for being tops in sales. Bolstering Ford's 1966 sales numbers was the hot-selling Mustang, with just under a half-million sold. Ford's Mustang had something for everyone, but Chevrolet had only rumors of a new pony car, code named Panther.

In 1966, the first federally mandated safety regulations came into effect. These changes were not yet dictated by law, but were requirements of the U.S. government's General Services Administration (GSA) for its fleet cars. If you wanted to sell cars to your biggest single fleet customer, you had to meet the GSA's rules. It was a way to implement safety standards before the NHSTA (National Highway Safety Transport Association) and FMVSS (Federal Motor Vehicle Safety Standards) mandates were written.

The Chevy II was completely restyled for 1966. Its new unitized body was very similar in external dimensions to the first-generation Chevy II. It featured a half fastback roofline that blended down to a mildly prominent quarter panel similar to the full-size Chevy. The new model appeared much larger due to its low profile and horizontal character line that ran along the side. *GM Media Archives*

With the exception of some minor ornamentation changes, the exterior of the 1966 Corvette appeared the same as the 1965. One of those changes was a newly styled full wheel cover. Federal laws ensured that 1966 would be the last year for any kind of protruding wheel spinners like the ones on this Sting Ray.

Most of the requirements made sense and should have been in place years before, especially with respect to the interior. In the 1950s, the instrument panel looked as if the designers had placed a grille inside. It was covered with diecast chrome nacelles and sharp edges—hardly the thing to be thrown against in an accident. Don Schwarz, chief designer of Chevrolet's interior studio from 1962 to 1967, said of the 1950s designs, "It was 'school's out.' Constant newness was what we were trying to achieve. When the standards hit, it threw cold water in our faces. We were forced to think more like engineers and less like stylists."

The nature of the industrial designer is to adjust to the mandates. Many changes were foreseen prior to the government's new requirements. Steve McDaniel was in charge of all GM interiors in the early 1960s. He decided to darken all of the instrument panels to reduce glare and directed young interior designer Blaine Jenkins to do so. "I told him it wouldn't match and would look really bad," Jenkins said. "He put his finger in my chest and said with his Georgia drawl 'If we don't do this stuff ourselves, were gonna have the government tellin' us!'"

It was a prophetic statement, but the changes put in place were more evolutionary than revolutionary. Chevrolet designers and engineers viewed

state police films of gruesome auto accident scenes that gave the designers an indication of what happened during a collision and how they had to redesign the interior to protect the occupants. Knobs on the instrument panel were rounded and control locations were changed or recessed. All 1966 Chevys had as standard safety equipment an outside rearview mirror, padded sun visors, two-speed electric windshield wipers with a washer, back-up lights, front and rear seatbelts, and a shatter-resistant inside rearview mirror. Padded instrument panels also became standard on all cars for 1966. Most instrument panel pad designs blended well, but the one on the Corvair perched on top of the instrument panel like a beanie. Chevrolet engineers were diligently working on numerous additional federally mandated changes that were scheduled to be in place for the 1967 models.

Also on engineer's drawing boards' were plans to accommodate changing emission laws. California (10 percent of the U.S. car market) led the way and demanded certain control devices. All Chevrolets sold in California, with the exception of the 425-horsepower 427-cubic-inch and 90-horsepower 153-cubic-inch L4, had an Air Injection Reactor System (AIR). The system consisted of a belt-driven pump and air-injection tubes. The tubes (one for each cylin-

Corvettes with 427 engines could be identified by the bulge in the hood and by the unique 427 flags on the front fender. Corvettes with 427 engines were also identified as the fastest cars Chevrolet sold in 1966.

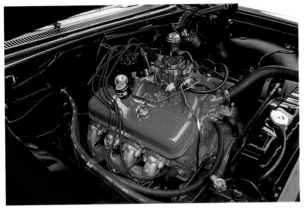

In 1966, three optional big-block engines were available for the full-size Chevy. The 396-cubic-inch, 325-horsepower engine was a carryover from 1965. Two new 427-cubic-inch engines were available, a 390-horsepower version (shown here) and a high-performance version rated at 425 horsepower.

Passenger safety was emerging as an issue of increasing importance. Federal regulators and auto manufacturers were working to protect passengers through better design. All 1966 Chevys, including this Super Sport convertible, had as standard equipment an outside rearview mirror, padded sun visors, a padded instrument panel, two-speed electric wipers, back-up lights, front and rear seatbelts, and a shatter-resistant inside rearview mirror.

der) were ported to the exhaust manifold. Fresh air was pumped into the manifold, intensifying and prolonging the burning action. Unburned hydrocarbons and some carbon monoxide were consumed.

Chevrolet struggled in 1966. Being outsold by rival Ford was tough to take. Almost every automotive enthusiast magazine in 1966 had a feature on Ford's NHRA or NASCAR racing efforts, and these articles were punctuated with double-page ads for high-performance items like Weber induction kits or Cobra cylinder heads. With the exception of the SS396 Chevelle, Chevrolet seemed reluctant to advertise its performance capabilities. Corvair was in a fast downward spiral with little hope of pulling out, and the Chevy II, even with its new attractive body, fell short of expectations. The bright spots were the Caprice and SS396 Chevelle. They told America that Chevy was still in the game and willing to play.

Chevelle

The 1966 Chevelle took on a new look. It was completely restyled and featured a fastback roofline on the two-door hardtops, similar to that of the Impala. The overall design of the Chevelle now resembled a scaled-down full-size Chevy. The tops of the quarter panels had a slight flare to them, leading to an end cap that housed a recessed taillight. A character line extended along the side of the body and washed out into the front of the hood, and the leading edge of the front fenders was cut back, giving the Chevelle the same forward lean as the Impala. The rear window on the two-door hardtop was countersunk. It was a new design theme that featured sail panels that extended rearward, giving the side view a fastback look. The only major deviation from the Impala design theme was the full front bumper.

In 1966, there was only one Super Sport Chevelle and it was a full combat model. Three-hundred-ninety-six cubic inches of big-block filled the engine compartment. The SS396 was available in three horsepower ratings starting at 325, a midrange engine at 360, and a top performer of 375 horsepower. The SS396 exterior was understated, with minimal ornamentation.

To reduce the overall cost, the 1966 SS was stripped bare of the expensive extras that were

The 1966 Chevrolet used the basic body of the 1965 Chevy and added new front-end sheet metal. The grille was the traditional Chevy egg-crate design, but with a tighter grid than on the 1965. This Danube Blue 1966 Impala Super Sport is equipped with optional full wire wheel covers.

New, more-upright rear fender end caps were used on the 1966 Chevy. This change dictated a new deck lid to match. Impala taillights were now a three-segment wraparound design, replacing the circular design, a standard since 1958.

added to the 1965 Z-16 model. The basic SS396 Chevelle was available in either a two-door hardtop or a convertible. The base engine was the 325-horsepower 396. It was backed up by a floor-shifted, fully synchronized three-speed manual transmission built by Borg Warner. A heavy-duty 12-bolt differential was standard, but positraction was not. The basic front and rear suspension was standard Chevelle with the addition of heavy-duty springs and shocks. Six-inch-wide steel wheel rims were standard, mounting red-line tires. Early production cars were fitted with small, dog-dish hubcaps, and full wheel covers were optional. A midyear change dropped the small hubcap, and the full wheel cover became standard.

The exterior of the SS396 Chevelle was trimmed modestly. Super Sport chrome script dressed the quarter panels. The 396 flags were prominently mounted on the front fender. Small SS396 badges were located in the center of the blacked-out grille and on the rear panel between the taillights. The most distinctive exterior feature on the SS396 Chevelle was its twin hood vents. These nonfunctional chrome fixtures were similar in appearance to the vents on the big-block Corvette hood, and the Chevelle's vents looked awkward compared to the GTO's smoothly styled hood scoop.

The standard interior of the SS396 Chevelle was austere. A bench seat was standard, available only in solid-color vinyl trim. Basic instrumentation was minimal, and even the tach was an option.

The option list for the SS396 Chevelle was lengthy and most buyers took advantage of it to fully dress-out their new musclecar. Power was on the minds of most buyers, and the optional 360-horse-power engine was similar to the 325-horsepower engine with the exception of a longer-duration cam. A 375-horsepower 396 powerhouse was on the option list, but few Chevelles equipped with this engine ever rolled through a dealer's showroom. These were available on special order only and took a long time to acquire. Few SS396 Chevelles were delivered with the standard three-speed manual transmission, as most SSs were ordered with the Muncie four-speed manual available with either wide or close ratios. The only automatic transmission available was Chevrolet's two-speed Powerglide.

At the top of the option list for most buyers were bucket seats. Strato buckets were now the optional seat, and the new Strato bucket seat had a more appealing look than the previous bucket seat design, which was of Corvair Monza heritage. A chrome center console was available to those ordering bucket seats. Rounding out the interior were instrumentation packages, a tilt wheel, and a wood-grained steering wheel. It didn't take long to add over $1,000 in options to the low base price of $2,776 on the 1966 Chevelle SS396.

In the February 1966 issue of *Hot Rod* magazine, Eric Dahlquist shared his thoughts on the new four-speed, 360-horsepower SS396 that Chevrolet had given him to evaluate. "As a synopsis of the random reflections that ran through our mind as we returned the car to the Chevrolet zone office, it could be said this SS396 was the type of vehicle we hated to part with. It has just the right measures of ride-handling and acceleration that would make it the nuts for all kinds of long trips. It's a fun car for today's dull traffic, and if it helps relieve tedium of travel, you can't ask much more."

Two of the hottest-selling coupes in 1966 were the new Caprice and the Chevelle SS396. As Chevrolet's full-size car was changing into a luxury car, the Chevelle was changing into a musclecar. *GM Media Archives*

Chevy's SS396 filled the vacant performance niche and re-established Chevrolet's performance image. The sales of 72,272 Chevelle Super Sports probably took more than a few sales away from Impala Super Sport models. One purchaser of a new 1966 SS396 Chevelle said, "I saw a white SS396 at an auto show, and I fell in love with it. It was the right size, and I could add the options I wanted. The Impala SS was just too big."

Chevrolet

The restyled 1965 Chevrolet had been a total success. No sane person would have suggested that the design could be improved upon. In the case of the 1966 Chevrolet, however, the new design was indeed better, and it didn't take much to do it.

The basic body was carried over. New end caps were added to the rear quarter panels giving a more upright, formal look, and to match the end caps, the rear of the deck lid was also more upright. Circular taillights, first seen in 1958, were deleted in favor of a three-segmented horizontal taillight that wrapped around to the edge of the quarter panel below the end cap. The horizontal line that continued to the leading edge on the 1965's front fenders was turned down on the 1966, creating a more precise look. The front was

also cleaned up with a tighter grid in the egg-crate grille.

Replacing the Super Sport as flagship of the 1966 Chevrolet line was the Caprice. Released as an option to the Impala in 1965, it was now a unique series. Chevrolet's strategy was changing with the arrival of the pony cars. Performance buyers wanted a smaller package with lots of image. Chevrolet was now free to explore the upper end of the scale with luxury never before offered in a Chevy.

The Caprice's success as a four-door in 1965 bred a new Caprice two-door coupe and two four-door station wagon models. Chevrolet designers placed their efforts in developing a new, less sporty upper body for the coupe. The long, flowing C-pillars of the standard coupe were shortened and a more upright quarter window was installed. The Caprice coupe had a taste of Cadillac with its more dignified look. Its design lent itself well to the addition of a vinyl top option, which pleased the marketing department, always looking for another option to sell.

More money was approved for the Caprice interior. Two-tone seat trim was added for a more upscale look. The new Caprice interior was available in both cloth or vinyl. Padding was added to the door trim panels to give them depth and a richer look. The Strato Bucket seat, the standard bucket

The 1965 Chevelle Z-16 paved the way for the 1966 Chevelle SS396. The SS396 was a performance machine stripped bare of the many extras added to the Z-16. Chevrolet made a few changes to the SS396 between the time this photo was taken and when production models reached the showroom. These changes included hood grilles, a blacked-out grille with an SS396 emblem, new wheel covers, and Super Sport quarter panel script. *GM Media Archives*

seat for the full-sized Olds, Pontiac, and Buick, was installed in the Caprice Coupe. The Coupes were also given a tunnel console and a four-gauge instrument cluster between the tunnel and instrument panel. The standard engine for the Caprice was the 195-horsepower, 283-cubic-inch V-8.

Under the beautifully styled body, the mechanical components of the 1966 Chevy were almost identical to those of the 1965, but a new 327-cubic-inch engine was available, rated at 275 horsepower. It replaced the two 327s (rated at 250 and 300 horsepower) offered in 1965. Two new 427-cubic-inch big-block engines were offered in 1966, with the high-performance version rated at 425 horsepower and being very similar to the 425-horsepower 396 offered in 1965. A moderate-performance 427 was rated at 390 horsepower, and it was similar to the 325-horsepower 396 with the exception of a longer-duration camshaft and the additional cubes. Those extra cubic inches were gained by increasing the bore on the 396 to 4.25 inches. The Turbo-Hyrdra-matic introduced in 1965 returned as an option only on the 325- and 390-horsepower big-block engines.

Chevy II

Rumors of the Chevy II's demise had been circulating in 1965 and Chevrolet executives did their best to defend what was a very well-engineered but poor-selling car. The design was dated and Chevrolet couldn't decide how to position the Chevy II in

Chevrolet was no longer reluctant about advertising performance cars. The four-gauge cluster at the top of the console was only available in the Super Sport or Caprice if bucket seats were ordered.

Zora Arkus-Duntov

Long before Zora Arkus-Duntov joined Chevrolet, he had already made his mark on the automotive world. Born in Belgium of Russian parents, he graduated from the Institute of Charlottenburg, Berlin, in 1934 as a mechanical engineer. Some of his earliest engineering work was on supercharging systems and diesel engines. He also had his engineering hand in the design of tractors, locomotives, and aircraft. During this time he entertained himself by racing cars and motorcycles, and it was while racing a flathead Ford engine that the idea for the Ardun cylinder head took shape. The Ardun head was a hemispherical design adapted to the flathead motor which substantially increased its horsepower. Duntov came to America in the early 1940s as a diesel engine consultant. He was encouraged by his contemporaries to go to Detroit, but he didn't think the engineers in Detroit would be interested in his ideas.

In 1953, Duntov attended a GM car show where he saw the first Corvette. He felt the car had a lot of potential and wrote GM. His evaluation of the car reached Ed Cole, Chevrolet's chief engineer at the time, who was impressed and asked Duntov to join Chevrolet as a research and development engineer. One of the first projects to which Duntov was assigned was the development of a passenger car fuel injection system. Duntov's racing experience landed him in the driver's seat of a preproduction 1956 Chevy for the Pikes Peak hill climb. His stock Chevy not only won its class, but set a new record that stood for 13 years. Fresh from the mountains of Colorado, Duntov proposed to set the Flying Mile record on the beach at Daytona in a new 1956 Corvette. A speed of 150 miles per hour was needed to set the record. The Corvette's 265-cubic-inch engine lacked about 30 horsepower, and Duntov suggested a new, higher-revving camshaft. The Duntov camshaft was born, generating the power needed to hit a record speed of 150.583.

In August 1957, Duntov was placed in charge of the Corvette program as engineering coordinator. He

Zora Arkus-Duntov

also was director of high-performance vehicles until 1964. In 1963, he assumed design responsibility for the Corvette engine and chassis, and in 1968, he was named chief engineer of the Corvette, the car that first drew him to Chevrolet. Duntov remained as Corvette's chief engineer until his retirement in 1975. Until his death in 1996, Duntov was a frequent and welcomed guest at Corvette events across the country.

While Duntov could never take credit for the invention of the Corvette, he certainly could take credit for its development into a world-class sports car. Duntov was a legend in his own time, something very few people can truly claim.

---

the market. Models and options like the Super Sport and convertible were added and then, just as quickly, taken away.

The 1966 Chevy II was completely restyled with an attractive new body. It was much sleeker than the previous design and looked much larger than it actually was. The 1966 Chevy II retained the 110-inch wheelbase and 183-inch overall length, but the

height was reduced by almost half an inch.

Other than the nameplates, the only other identifying features carried over from 1965 on the new Chevy II were the two large headlights. The sides of the body were almost flat except for a raised horizontal bead that ran along the side. The tops of the front fenders had a slight peak to them that rounded over to the chrome headlight bezel, and

Plain Jane it ain't

Chevy II-Styled The Chevrolet Way

The Chevy II was completely restyled for the 1966 model year. The new body had the look of a much larger and more expensive car. The tops of the quarter panels had a slight flare similar to the shape on the full-size Chevy. The 350-horsepower 327-cubic-inch V-8 was the highest horsepower engine available in the new Chevy II.

Corvette bodies were pieced together much like a plastic model car kit. The slightly darker line running forward from where the worker in the foreground is sanding is one of the body joints that required hand-finishing prior to paint. Corvette body integrity was exceptionally high in 1966. *GM Media Archives*

the grille was a simple anodized aluminum egg-crate design. The roofline was a semi-fastback design, not as tapered as the Impala or Chevelle, and the thick C-pillar gave the roof a more formal look. The tops of the quarter panels transitioned to a flat area across the rear at the top of the taillight. The taillights were rectangular and stood vertically, giving the suggestion of a more expensive car. A large rear bumper, contoured to the shape of the body, matched the one on the front in size and style. It was a well-balanced design with the two-door hardtop the most attractive of the group.

"We couldn't get the guys revved up because we weren't revved up ourselves," Chuck Jordan recalls about the design of the 1966 Chevy II. "It doesn't have much feeling or emotion. I wonder why we couldn't have done better?" Mom-and-pop projects like the Chevy II, where cost is a big consideration, were difficult to work on compared to a project like the Corvette. When asked about the 1966 Chevy II,

one former Chevrolet designer admitted, "I really don't remember anything about that car."

Missed or dismissed by most performance enthusiasts in 1966 was the 350-horsepower engine option available in the Chevy II. Backed by a Muncie four-speed transmission, it was a true musclecar. Its power-to-weight ratio was better than the SS396 Chevelle equipped with the 360-horsepower engine, and when road tested, it outperformed its midsize sibling by a large margin. *Car Life* magazine tested a 350-horsepower Nova SS for its May 1966 issue, and the editors raved about the performance of the engine and balance of the chassis. "Unlike some samples from the super car spectrum, it maintains a gentleness along with its fierce performance potential; its power/weight ratio is second to none and it is definitely better balanced than most."

Chevy II sales in 1966 reached 172,485, bettering the 1965 output by 50,000, but it was still trailing Ford's Falcon in sales. The Chevy II lacked a lot of extra-cost options, which were reserved for the Chevelle with hopes that a potential Chevy II buyer would trade up to the next-in-line, more-expensive model.

Corvette

Technical innovations and horsepower increases characterized the 1965 Corvette. Little was left to do in 1966 but refine those features. Corporate politics relaxed just long enough for Duntov's engineers to add the missing 30 cubic inches, and two 427-cubic-inch engines were available for the 1966 Corvette, one rated at 425 horsepower with solid lifters, the other at 390 horsepower with hydraulic lifters. The

The success of the upscale 1965 Caprice four-door made way for the Caprice Coupe in 1966. It received all the luxury appointments of the four-door Caprice and a new, more-upright roofline that set the tone for the Coupe's formal image. *GM Media Archives*

425-horsepower version was initially rated at 450 horsepower and a few were sold with an air cleaner decal denoting that rating. Even though the 427s were more powerful and versatile than the 1965 396s, they didn't seem to excite the public. Most of the hype surrounding the big-block introduction had been spent the previous year.

The standard engine, a 300-horsepower 327 cubic inch, was at the lower end of the horsepower scale. The other small-block available in 1966 was the 350-horsepower 327. The 300-horsepower and 390-horsepower were the only engines available with Chevrolet's Powerglide automatic transmission.

The exterior of the Corvette changed very little from the 1965. Three large vertical louvers were still prominent on the front fenders, the vents on the coupe's sail panels disappeared, and small, subtle ornamental changes were made.

Corvair

Corvair sales in 1966 were dismal at best. Total production of 103,745 units was less than half that of 1965. The public's perception of the Corvair was irreparably damaged. Only small engineering and safety changes were made for the 1966 model year, and the Corvair line-up remained the same with the Corsa as the top model. The handwriting was on the wall for Chevy's sporty little car.

The mating of body and chassis during assembly is a delicate operation performed by several employees and a large hydraulically operated body fixture. The employee on the right is positioning a fan shroud on the front of the engine, while the employee to the left of center holds the hydraulic controls for the fixture. *GM Media Archives*

Chapter 8

1967 The Hot New Camaro

Many changes affected the auto industry in 1967. New federally mandated safety features were required on all passenger cars and more regulations were forthcoming. Chevrolet was launching its new pony car, the Camaro, and wondering what would become of the Corvair.

To comply with federally mandated standards, 29 new or improved safety features were added to all Chevrolet passenger cars. General Motors had been working on an energy-absorbing steering column for five years. The hazard had been that of the driver being thrown forward during a frontal impact while the steering column was moving rearward. GM's new column absorbed the force by collapsing, thereby shortening its overall length. The column was also designed not to extend into the passenger compartment.

New door lock mechanisms were designed and installed to prevent the car doors from opening upon impact during an accident. Four-way hazard warning switches were installed on all Chevrolet steering columns, and seatback latches were added to the folding front seatbacks. These latches were installed to prevent the rear-seat passenger from being catapulted into the front seat during a crash.

Chevrolet disc brakes were first seen on the 1965 Corvette due to its small production run. Chevy engineers needed experience with the new

This early Camaro prototype carried most of the elements of the final design with a few exceptions, such as the horizontal body pinstripe and parking lights in the lower valence panel. *GM Media Archives*

In 1967, Chevrolet introduced the SS427, a full-size equivalent to the SS396 Chevelle. It featured special ornamentation and a standard 385-horsepower 427-cubic-inch engine under a domed hood. The parchment vinyl Strato-bucket seats inside this SS427 are fitted with optional headrests. *Paul Zazarine/Muscle Car Review*

technology and the Corvette was their test mule. They felt uneasy offering a feature on a high-volume standard passenger car without some real-world experience. By 1967, Chevrolet was ready to offer disc brakes as an option on all car lines except Corvair, whose drum brakes provided adequate stopping power. Standard with the disc brake option were attractive rally wheels. These silver-painted wheels were made of stamped steel and had five slots in the center web section. A small hubcap fit over the lugs and a trim ring was added to the outer rim. These wheels were a simple design that is still attractive today.

Also added to the braking system of every car was a dual master cylinder. It split the braking into pairs of wheels, providing a degree of backup in the case of certain brake failures. A warning light was also installed on the instrument panel to let the driver know of a possible brake system problem.

In an effort to protect pedestrians, the government limited projections on the exterior of the car. Corvette's tri-bar knock-off wheel nut was deemed to have the same effect as the spikes on wheels of Roman chariots, so they were removed in favor of a bolt-on wheel of the same design. Other safety related items standard on all 1967 Chevys included two-speed windshield wipers with washers, back-up lights, padded sun visors, rear seatbelts, an exterior rearview mirror, and an antiglare interior rearview mirror. The sharp edges of the interior mirror were covered to protect passengers in case of accident. Exterior mirrors and ornamentation were also redesigned for pedestrian safety. The era of vehicle safety and government regulation was just beginning. Within a few years, the design, engineering, and testing of the American passenger car would change dramatically.

In 1957, Chevrolet offered warranty coverage on its new cars for 90 days or 4,000 miles. Such a short warranty did little to instill confidence in the new buyer. By 1967, the warranty had been extended to two years or 24,000 miles for the original and subsequent buyers. An extended warranty of five years or 50,000 miles also covered the powertrain, steering, suspension, and wheels. In 1967, Chevrolet buyers were definitely getting their money's worth.

Engineering delays on what would eventually be the 1968 Corvette resulted in another year of the original Sting Ray body style. Only two 327-cubic-inch small-block engines were available in the 1967 Corvette, rated at 300 and 350 horsepower.

The 1967 Corvette carried the same lines as the original 1962 Sting Ray. Most of the changes were on the front fender, namely the redesigned front fender louvers and the deletion of the crossed flag emblems. Rally wheels with trim rings and hubcaps were standard and a bolt-on version of the knock-off aluminum wheel was optional. This Marina Blue 1967 convertible is equipped with optional side pipes and red-line tires

The entire auto industry was down slightly in 1967 with only 7.6 million units sold, the lowest level since 1963. This low level of overall production was due in part to a crippling UAW strike against Ford Motor Company. The good news was that Chevrolet's segment of the market in 1967 was just over 26 percent. Even better, it outsold Ford.

General Motors was doing its best to hold down the base price of its cars, as can be seen from this quote from the annual report: "Prices for comparably equipped GM cars today are the same as or lower than the 1959 models introduced in the fall of 1958. Based upon comparable products, the U.S. Bureau of Labor Statistics index for new car prices for 1967 model passenger cars shows a price decrease of 8 percent from 1959 models. Over the same period the U.S. Consumer Price Index has risen 14 percent." While the base prices were still relatively low, the cost of upscale models and added options pushed sticker prices over the $5,000 mark.

On December 15, 1967, General Motors entered into an agreement with the UAW for a new three-year contract. The agreement followed those previously settled with Ford and Chrysler. The highest paid hourly workers in the industrialized world were getting automatic raises of 20 cents per hour, plus an additional 30 cents per hour for skilled trades. Guaranteed in 1968 and 1969 were raises of 9 to 18 cents per hour each year, depending on the employee's base rate. These raises, added to an additional paid holiday in 1968 and 1969, turned out to be an expensive agreement for GM. Also costly were the federally mandated changes required on all cars, as additional required equipment had to be engineered, designed, and installed.

In its haste to protect the American citizen, the U.S. government was writing regulations for vehicles without knowing if they were valid or could even be met. It hired a few former Detroit auto engineers to develop these standards, but the government failed to fund these people for testing the rules they were writing. The burden fell upon the auto makers to meet the new dictates, and the government expected the auto companies to come back with recommendations, at which point the regulation could be rewritten. This ongoing discussion over the government's requirements gave the impression the auto makers were soft on issues of safety. Following the disastrous press coverage Chevrolet had received over the Corvair, it had to make certain not to get another black eye on safety issues. Chevrolet diligently tried to implement all mandates, but auto safety issues were a problem for both auto makers and government officials.

The lines of the 1967 Chevrolet were much more crisp than those on the 1966. The distinctive flared edge at the top of the quarter panel was retained, but the wheel openings were changed to an elliptical shape. The Impala Sport Coupe's fastback roof ran to the edge of the deck lid without a filler panel in between. *GM Media Archives*

Camaro

Chevrolet General Manager Pete Estes, at a press luncheon in New York for the International Auto Show in May 1966, was asked if Chevrolet would have a Mustang-type car in 1967 and what it would be called. Before he could answer, another reporter yelled, "I didn't hear the question." Estes said with a laugh, "He wants to know the name of the Panther." Estes went on to say, "It's the sporty car market that our new car is aimed at. I don't know how many sales we're going to have, but we'll be competitive, you can bet on that."

The Panther was renamed Camaro and went on to sell 220,917 units in 1967, about half of Ford's Mustang sales for the same year.

When Ford introduced the new Mustang in 1964, Chevrolet executives quietly watched the market, knowing their Corvair had been totally redesigned for 1965. Then two things happened. First, the press was in a feeding frenzy over Ralph Nader's book, which declared the Corvair unsafe. The bad publicity kept buyers away from Chevy showrooms. Second, the public loved the Mustang. It was a simple, sporty car with a conventional driveline and chassis. The Mustang offered a long option list, so each customer could build the car to suit his or her taste and pocketbook. The Corvair was too exotic for the taste of the average car buyer. The solution was to build a simple sporty car, the Camaro.

Bill Mitchell loved sporty cars, cars with curvy lines, bucket seats, a performance engine, a floorshift, and a maximum seating capacity of four. It was always his desire to build a four-place sporty car, and it seems as though he always had someone in one of the studios working on such a design. The Corvair Monza had been conceived in this manner, but the Corvair was on its way out. To compete

With the addition of the new Camaro, the 1967 Chevrolet car line grew to seven distinct models. From left to right: Corvair, Corvette, Chevy II, Chevelle, Impala, and in the foreground, Camaro. *GM Media Archives*

against the Mustang was Mitchell's chance to design a breed of car he favored.

Mitchell envisioned the Camaro as a four-passenger Corvette. It wouldn't have the Mustang's angular proportions—squared-off top, short rear deck, and long front. The Camaro would have voluptuous curves with a more rounded look. "The guys in the studio were fired up with the opportunity to do the Camaro. It was a whole new animal," Chuck Jordan recalled. "People imagine these ideas, and if you're revved up and excited about a project, these things start to come out. You wake up at night with ideas; you don't wake up and think about Chevy IIs."

Under the direction of Henry Haga, Chevrolet's Studio No. 2 created the Camaro's design. The Camaro's inspiration came from a show car called the Super Nova, displayed at the 1964 New York Auto Show (although as early as 1962, lines for a Camaro-type car had been laid down). It emphasized roundness and fluidity, typical of mid-1960s Chevrolet design. This same design philosophy was exhibited in the 1965 Corvair.

In the book *Camaro Style, Speed, and Spirit*, Haga spoke about the design of the Camaro: "We felt very strongly about reducing design to its simplest form, using only one peak down each body side, interrupted by accented wheel arches. The profile of the car also was very simple, using the classic approach of crowned fender lines, with their high points directly above the accented wheel arches. We purposefully avoided any contrived design lines and superfluous detail. Even the execution of the wide, horizontal loop front end and grille, with its hidden headlamps in the Rally Sport variant, was as pure in concept as we could make it."

The horizontal blade bumpers fore and aft restated the theme of simplicity and gave the illusion of added width. The headlights were single units placed in the corners of the oval grille. The taillight and back-up light assemblies were simple, two-segment horizontal units.

The interior package dimensions for the Camaro were very similar to those of the Mustang. The overall length of 184.6 inches was 1.3 inches longer than the Mustang, and the Camaro was a fraction of an inch lower and wider than the Mustang. The Camaro was to be built only in two body styles, a coupe and a convertible. A two-seat convertible and a fastback comparable to Mustang's 2+2 were proposed, but never approved.

Just over 100,000 Chevy IIs were sold in 1967, the lowest sales figures in its history. The 1967 model was almost identical to the 1966, with minor trim changes. The powerful 350-horsepower engine was dropped, leaving the 275-horsepower engine as the top performer in 1967. *GM Media Archives*

The Camaro's instrument panel looked like that of the next-generation Corvette, with two large circular instruments on a receding surface in front of the driver. Controls were moved up on the instrument panel and rounded off to protect the driver in case of an accident. The center of the instrument panel housed the heater and radio controls, both recessed for safety. This center area blended down into a console that split the two front bucket seats. Rear seats were no larger than those of the Mustang and were not suitable for anyone over 12 years of age. Foam-backed molded door panels, introduced on the 1965 Corvettes, were being used for the first time on a Chevrolet production car in the Camaro's deluxe interior.

It was determined that many of the Camaro's basic components would be off-the-shelf technology in an effort to reduce engineering time and cost. The decision to use the Chevy II's cowl restricted the overall design. Its height raised the center of the car more than Mitchell and his designers wanted.

The Camaro's body was a unitized design with the addition of a front subframe. Subframes had been used on European cars with a high degree of success. GM's first use of a subframe was on the 1966 front-wheel-drive Olds Toronado. This was Chevrolet's first effort, and it was highly effective. The subframe carried the engine and front suspension components. It mounted to the body with rubber biscuits that blocked out noise and vibration.

The 1967 Chevelle was updated with new front sheet metal and new rear fender caps featuring wraparound taillights. A special domed hood was part of the SS396 package, along with a blacked-out grille. *GM Media Archives*

This same subframe would show up under the front of the 1968 Nova.

Budget constraints limited the design of an effective rear suspension for the Camaro. The Chevy II axle and single-leaf springs were used and were marginal performers when V-8 power was applied. Traction bars were added in an attempt to lessen rear wheel hop.

In 1967, a bare bones Camaro coupe sold for just under $2,500. The customer could choose from

Pete Estes

On July 1, 1965, E. M. "Pete" Estes succeeded Bunkie Knudsen as general manager of Chevrolet Division. Like Knudsen, Estes earned his stripes as the head of Pontiac for three and a half years prior to his move to Chevrolet. Estes championed Pontiac's performance image and increased market share from 6.4 percent to 9.5 percent, moving Pontiac up to number three in sales.

Estes was born in Mendon, Michigan, on January 16, 1916, and as a young man he wanted to be a civil engineer. In the summer of 1934, while waiting to enroll at Michigan State College, he read a newspaper story about the opportunities at General Motors Institute in Flint, Michigan. He applied and was accepted, beginning his GM career on October 1, 1934. Following graduation, Estes worked at the GM Research Laboratories with famed GM inventor C. F. Kettering.

In 1946, Estes joined the Oldsmobile division in Lansing, Michigan. There he worked as a motor development engineer with the Kettering team on the industry's first high-compression engine. In 1954, Estes was promoted to assistant chief engineer of Oldsmobile in charge of body design, chassis design, and engineering standards.

On September 1, 1956, he was promoted to chief engineer of Pontiac Motor Division, working under General Manager Bunkie Knudsen. Estes and his engineering team at Pontiac were responsible for the successful wide-track Pontiac chassis. They also developed the Tempest with its unique front engine and rear transmission. In 1961, when Knudsen moved to Chevrolet, Estes filled his position at Pontiac.

His three and a half years at Pontiac were impressive, as Pontiac sales doubled during his tenure as general manager. While other GM divisions were pruning performance cars from their line-up, Estes championed Pontiac's GTO.

As general manager of the Chevrolet Division, he was immersed in the design and engineering of Chevrolet's

Elliot "Pete" Estes

sixth line of cars, the new 1967 Camaro. For 1968, Estes directed the largest single-year model change in the history of Chevrolet. The Chevelle, Chevy II Nova, and Corvette were totally redesigned, while significant styling, safety, and engineering upgrades were made to the balance of the Chevrolet line. During Pete Estes' reign, Chevrolet produced the largest selection of high-performance cars in its history.

a long list of factory and dealer options equaling that of the Mustang. Most Camaro options were within pennies of those on a Mustang.

The standard engine for the Camaro was the 140-horsepower six-cylinder engine from the Chevelle. An optional six-cylinder engine was the 155-horsepower 250-cubic-inch engine from the full-size Chevy. Both of these engines were standard with a three-speed manual transmission and available with the Powerglide automatic.

Chevrolet went right for the throat with its selection of V-8 engine options in the 1967 Camaro. The smallest V-8 was a two-barrel carbureted 327-cubic-inch engine rated at 210 horsepower. The addition of a four-barrel carburetor and increase in the compression ratio to 10.0:1 jumped the horsepower rating to 275. Available only on the Camaro in 1967 was the new 350-cubic-inch small-block rated at 295 horsepower. Some 0.25 of an inch of stroke was added to the 327's crankshaft to gain the extra cubic inches.

Corvair sales spiraled downward in 1967. Slightly more than 27,000 units were sold in 1967, one-third of what was sold in 1966. The turbo was discontinued and this Monza Sport Sedan was one of only five Corvair models available. *GM Media Archives*

To counter Mustang's 390-cubic-inch V-8, Chevrolet added the powerful 396-cubic-inch engine to the Camaro. It was available in two horse-power ratings, a 325-horsepower hydraulic-lifter version that was shared with the Chevelle and full-size cars and a 375-horsepower version, optional on the Chevelle. All Camaro engines were standard with a three-speed manual transmission and available with an optional four-speed manual. There were two automatic transmissions available, a Pow-erglide for the six-cylinder engines and small-block V-8s and an Turbo-Hydra-matic for the 325-horse-power 396.

One Camaro option that stirred a lot of interest in 1967 was the Z/28. The Z/28 was the brainchild of Vince Piggins, Chevrolet's manager of product promotion engineering (that is to say, race program manager in a nonracing company). Piggins pro-moted the Z/28 on the basis of increasing sales, but in his mind that meant winning races. The Z/28 was

configured to compete in the SCCA (Sports Car Club of America) Trans Am series.

The heart of the Z/28 was its high-performance 302-cubic-inch V-8. Chevrolet took a basic 327 block with its 4-inch bore and added a crankshaft with a 3-inch stroke. Chevrolet's engine group, knowing the potential of this combination, had been experimenting with this configuration as early as 1958. A high-rise aluminum manifold bor-rowed from the Corvette was topped with a large Holley four-barrel carburetor. Big-valve Corvette heads and a long-duration solid-lifter camshaft were added to build an engine that developed 290 horsepower at 5,800 rpm. Backing that high-revving V-8 was a close-ratio four-speed and stan-dard 3.73 rear axle. Chevrolet added a heavy-duty suspension and some racing stripes to the coupe body, and the Z/28 was born. Only 602 were built in 1967, and they had their teething problems against the Mustangs in competition. Lessons

Corvette coupes were still popular in 1967. This particular coupe is powered by one of the optional 427-cubic-inch engines. All 427-optioned Corvettes were equipped with the scooped hood shown here. *GM Media Archives*

The 1967 Camaro was originally developed under the code name Panther. It was only available in two two-door body styles, a coupe, and a convertible. Its physical dimensions were similar to Ford's Mustang. The Camaro was constructed with a unitized body mated to a front subframe. *GM Media Archives*

learned on the track were fed back to Chevrolet's engineering staff, and resultant changes benefited not only the Z/28 but all Camaros.

Corvette

The new Corvette body style based on the Mako Shark II show car was rumored to be released in 1967. Due to the extra engineering needed for the new body structure, the release was delayed until 1968. What the 1967 Corvette customer got was probably the best of the Sting Rays. Chevrolet had four years to perfect body construction techniques on the Sting Ray, making 1967 body quality exceptional. The most visible exterior change was the removal of three large front fender louvers. They were replaced with a set of four angular louver-like vents. The front fender no longer carried an emblem denoting engine size. Corvettes equipped with one of the three 427 engines had those numbers spelled out in chrome on the side of a new hood scoop. This new hood, with its contrasting color stinger, was exclusive to the 427s. The scoop was nonfunctional, but could be converted to provide fresh cool air to

the engine compartment. A notable change to the rear was the relocation and redesign of the back-up lights. The 1967 Corvette featured a single horizontal light above the license plate.

New 15-inch rally wheels were the base wheel rim, like those fitted to full-size Chevys with the disc brake option. These wheels replaced the standard steel wheel rim and full wheel cover. Knock-off aluminum wheels were no longer available due to legislation prohibiting protruding spinners. Instead, an aluminum wheel of the same basic design as the knock-off was offered. It attached with lug nuts and had a small center cap.

Under the hood, the same two 327-cubic-inch engines were available as in 1966, the base 300-horsepower and the optional 350-horsepower. Three different big-block engines were available in 1967: The 390-horsepower 427 returned as the most civilized of the group, and Chevrolet added three Holley two-barrel carburetors to that engine and the horsepower increased to 400. At the top end was the 435-horsepower, more commonly known by its option code, L71. The 435-horsepower engine was only available with a four-speed transmission, while the other two 427s could be ordered with a heavy-duty Powerglide.

Corvettes sold well (22,940) in 1967, even though the market was aware of a new design just around the corner. The Mako Shark II show car had been on display at auto shows, and its design was rumored to be the look for the new Corvette.

Chevrolet

The family resemblance between the Chevrolet, the Chevelle, and the Chevy II was very evident in 1967. Each model was distinctively styled and stood on its own, but subtle design cues gave each the look of the lineage—a tribute to Chevrolet's design staff. The full-size Chevy's new look was the result of a full face-lift for 1967. The new body had the same beautiful proportions as the previous two years with a few sharper edges. The flared quarter panels and fastback roofline on the two-door hardtops carried the look of the previous year. The new body appeared even longer than its predecessor but measured the same. The long fastback roof on the two-door hardtops extended to the deck lid without a filler panel in between. The wheel openings were enlarged slightly and had an elliptical opening. The bumpers fore and aft were full width, replacing the distinctive blade front bumper and lower grille carried by the 1965 and 1966 models.

Two distinct Impala Super Sport models were offered in 1967. The standard Super Sport featured the Chevrolet's usual bucket seat and trim upgrade to the Impala two-door hardtop or convertible. It was available with engines ranging from a six to a 396, while the SS427 was built around the 385-horsepower 427-cubic-inch engine, the same 427-cubic-inch engine offered in the Corvette with a 390-horsepower rating. With the SS427 Impala, Chevrolet was trying to rejuvenate the full-size SS option by offering a full-size car with the same performance and panache as the SS396 Chevelle. The SS427 was devoid of exterior ornamentation, except for a unique crossed flag emblem on the front fender with the numbers 427 above. In the center of the blacked-out grille and at the lower edge of the deck lid were SS427 emblems similar in look and placement to those on the SS396 Chevelle. Perhaps the most distinctive item on the SS427 Impala was the domed hood. The standard Impala's hood had a small peak down the center. The SS427 had a raised, tapered blister that ran down the center of the hood similar to the one on the small-block Corvette. Topping that blister, toward the rear, was a chrome insert that simulated three square carburetor intakes. Unfortunately, General Motors restricted all passenger cars to a single carburetor in 1967. Otherwise, the SS427 might have had the Corvette's tri-power under that special hood.

Topping Chevrolet's 1967 line of full-size cars was the Caprice, available in three models: a two-door hardtop, a four-door hardtop, and a station wagon. The two-door carried a more formal roofline, similar to that of the 1966 Caprice Coupe.

Camaros were available in several models and, like the Mustang, with a long list of options. This very male-oriented advertisement ran in the April 1967 issue of *Motor Trend* magazine. It featured the types of Camaros that could be ordered and the options available.

Imitation walnut trim and luxury cloth covered the interiors, and Strato bucket seats were optional on the coupe. Extra thick deep-twist nylon carpeting covered the floors as well as the lower portion of the doors. Standard on the Caprice were front fender lights, very distinctive lights that wrapped around the sides of the front fenders and lit up when the headlights or parking lights were on.

Chevelle

Following a very successful year, the 1967 Chevelle received a face-lift. The basic body was the same, but new front fenders, hood, and rear treatment were added. The new look gave the appearance of a slightly heavier car. The front fenders had the same forward lean as the 1966 Chevelle, but lost the wraparound grille. The grille area was opened slightly, and the anodized aluminum grille was Chevrolet's standard egg-crate pattern with more prominent horizontal bars giving the appearance of width. New end caps were fitted to the ends of the quarter panels, and a new taillight design wrapped around to the side.

The Chevelle's chassis and engine combinations were the same as in 1966, with the exception of the optional SS396 engine, now rated at 350 horsepower instead of 360. Also on the option list for 1967 was a 375-horsepower version of the 396. This engine was available on a limited basis and only with a four-speed transmission. It was the hot ticket for performance. *Motor Trend* magazine tested three combinations of 396-powered 1967 Chevelles and the hot 375-horsepower version turned in some impressive times. Its quarter-mile elapsed time was 14.9 seconds at 96 miles per hour; zero to 60 was a quick 6.5 seconds. These times were run in full street trim with F70-14 Wide Oval tires. Along with the engine's performance, the reviewers were ecstatic about the suspension. *Motor Trend* magazine's Robert Shilling wrote, "The SS396 is far and away the best handling and riding GM product we've driven in a long time."

The 1967 SS396 was a very reasonably priced performance car. The base two-door hardtop listed for $2,875.00. An upgrade to the 350-horsepower engine was only $105.35, and a four-speed transmission cost $105.35. Disc brakes were available for the first time on a Chevelle for only $79.00, including rally wheels. Also available for the first time with either the 325-horsepower or 350-horsepower engine was the Turbo 400 transmission. At $231.00, it was double the cost of the Powerglide, but worth every penny.

Chevy II

The 1967 Chevy II appeared identical to the 1966. The only changes made to the Chevy II were those dictated by law and the addition of the disc brake option. Minor trim changes were made, but to the untrained eye, the 1967 model looked the same as the 1966. Dropped from the engine option chart was the 350-horsepower 327. This left the 275-horsepower 327 engine at the top of the engine option list. The 275-horsepower 327 was a good engine, but it was not the performance equal of the 350.

Sales of the Chevy II in 1967 slumped to a disappointing 106,500, the worst in the car's history. The Chevy II probably suffered as much from the excitement of the new Camaro as anything else. People were also buying bigger cars with lots of options,

The 1967 Camaro was aimed directly at the youth market. To get the Camaro to market as quickly as possible and to reduce costs, many off-the-shelf components were used. Other elements, like the subframe, were designed to be interchangeable with future models. *GM Media Archives*

and Chevy was not alone in singing the small-car blues. Ford's Falcon sold only 64,335 units in 1967.

Corvair

Chevy's sporty little compact car was in for another year of decline in both sales and image. The 1967 Corvair line-up was trimmed to five models and the turbo engine option was no longer available. Corvair sales in 1966 totaled just over 100,000 units and in 1967 dropped to a dismal 27,253. A good percentage of the loss could be traced to the customer's selection of the Camaro over the Corvair and the rest to negative media attention. General Motors also made the unfortunate mistake of investigating Ralph Nader in an attempt to discredit him. The news of GM's investigation brought additional scorn upon the beleaguered company, so much so that GM President Jim Roche publicly apologized to Nader.

Even with falling sales, Chevrolet didn't cancel Corvair production for 1968. Chevy engineers went about their business upgrading the Corvair to meet new federal safety standards. One theory was that if Chevrolet canceled production in the midst of litigation, it would give critics reason to claim the car was unsafe. Still, 1967 was the last time anyone saw Chevrolet advertise the Corvair.

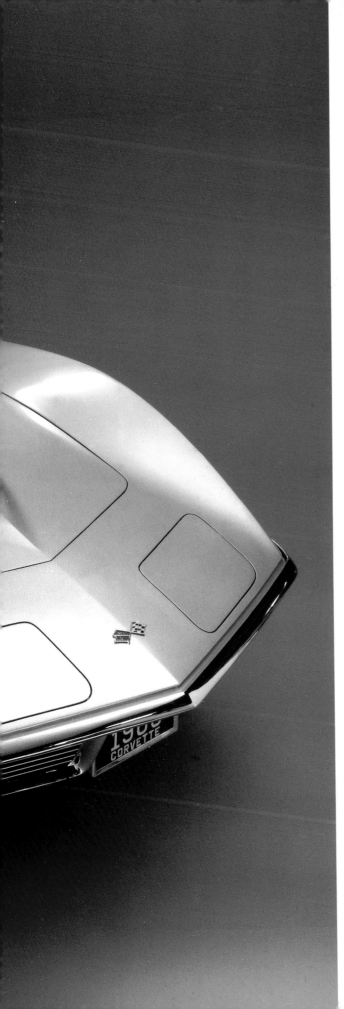

Chapter 9

1968 A Sexy New Corvette Debuts

Chevrolet's passenger car line-up in 1968 remained unchanged, offering six distinct car lines. The number of models within those lines was reduced from 89 to 82. In many cases, a discontinued model could be built with the proper selection from the extensive option list. The Corvette, the Chevelle, and the Chevy II were completely redesigned for 1968, and the full-size Chevy received new front sheet metal and a smartly restyled rear, while the Camaro was essentially the same for 1968 and the Corvair was left unchanged.

Although sales were good in 1967, profits were not what they should have been due to the increased cost of materials and labor. Boosting the cost of cars in 1968 was the addition of more federally mandated safety and emissions equipment.

Among newly mandated safety features for 1968 were four side-marker lights, which were illuminated whenever the parking lights or headlights were turned on. The two lights in the front had amber lenses, and lenses for the two lights in the rear were red. The general public scoffed at the addition of these lights, but they did help identify a car at night from the side.

The new Corvette coupe carried the design themes of the Mako Shark II into a production vehicle. The Corvette's long, low front end was only attainable with the integration of hidden headlights (under the rectangular panels). The panel between the windshield and cowl vent concealed the windshield wipers. When activated, the panel would pop up and move forward to allow the wipers to sweep the windshield. *GM Media Archives*

The new Corvette interior had the look of a jet airliner with its large center console and gauge package. Shoulder room was reduced in the new design and clearance between the woodgrain plastic steering wheel and the door panel was at a minimum. *Corvette Fever*

The new Corvette featured a Coke bottle-shaped body pinched in at the doors. The door release consisted of a flap on the top of the door covering a hand hold and a countersunk push-button release. Only the leading three of four large front fender louvers were functional, as the rear one was located on the cowl structure. *Corvette Fever*

Interiors received the most safety revisions. Front-seat shoulder harnesses were required, but because of the tangle of belts, most consumers left them clipped to the headliner. Door armrests were redesigned to prevent soft tissue damage as a result of a side impact. The small, blade-shaped chrome-plated plastic armrests Chevrolet used for years were replaced with a much deeper, thicker, and softer armrest. The depth of the armrest spread the impact load down to the hip area, thereby reducing the soft tissue damage. The new armrests also shielded the interior door handles.

Door hinges were now made of stamped steel instead of cast-iron. This allowed the hinges to deform in an impact, rather than break. Window regulator handles, door lock knobs, and ashtrays were redesigned to yield under a prescribed force. Dome lamps were now all plastic with no sharp corners. Textured, vinyl-coated foam padding was added to all windshield A-pillars.

Even something as seemingly innocent as the glovebox door lock was redesigned for passenger safety. The old design used a push-button to open the door. In an accident with a frontal impact, the inertia of the weight of the button was enough to release the latch, allowing the door to flop open. The sharp edge of the glovebox door could then do serious damage to

New exhaust emission equipment appeared by federal mandate on all 1968 models. This particular 325-horsepower 396-cubic-inch engine is equipped with GM's Controlled Combustion System (CCS). The CCS increased combustion efficiency through specific carburetor and timing calibrations. This system also includes a thermostatically controlled, closed-element air cleaner.

the lower extremities of a passenger thrown into the open door. The new door latch required a twist of the latch to open the door, a simple solution.

A big effort was made to reduce interior reflections from bright surfaces. Steering wheel trim was no longer highly reflective, smooth, bright chrome. A textured dull chrome surface was added to obscure reflections, and this same nonreflective finish was added to the gear selector lever, turn signal lever, automatic transmission quadrant housings, and rearview mirror back and mounting bracket. In an effort to reduce theft, all 1968 Chevrolets were equipped with a warning buzzer that alerted the driver if he opened the door with the key left in the ignition.

New exhaust emission equipment appeared on all models by federal mandate. All engines were now equipped with a 195-degree thermostat. Engineers found that hotter-running engines reduced hydrocarbon emissions, and two different systems were used by Chevrolet to control exhaust emission: the Controlled Combustion System (CCS) or the Air Injection Reactor System (AIR). The AIR system was similar to the one introduced in 1966 on Chevys with California emissions equipment. In this system, a belt-driven pump forces fresh air into the exhaust manifold to burn excess hydrocarbons. The AIR system was used on all Corvettes, Corvairs, and on other models equipped with manual transmissions. The CCS increased combustion efficiency through specific carburetor and timing calibrations, and also included thermostatically controlled closed element air cleaners.

These two systems were the first attempts to solve the emission problem for all 50 states. Not all systems worked flawlessly. Driveability suffered and mileage dropped. Many owners removed the foreign-looking equipment as soon as the warranty expired. Most performance enthusiasts removed the AIR system pump upon delivery, feeling it robbed them of horsepower.

A 307-cubic-inch V-8 replaced the 283-cubic-inch engine in the Chevrolet, Chevelle, and Chevy II. This new small-block V-8 kept the same bore as the 283 (3.825 inches) and used the stroke of the 327 (3.25 inches). The 307 was rated at 200 horsepower and was available only with a two-barrel carburetor. A new disposable oil filter was standard on all V-8s.

Lower hood profiles spurred an increase in the use of cross-flow radiators in 1968. The increased engine operating temperature also demanded upgraded cooling capacity. The optional heavy-duty radiator on the 1967 full-size Chevrolet became the standard radiator. Traditional carburetor-to-accelerator pedal rod and pivot linkage gave way to a new cable-controlled system on the Chevelle and Corvette with all engines. This new design was also used on the Camaro and Chevy II with the L-6 engine only. The new linkage gave engineers some design freedom, since engine compartments were getting tighter.

In 1968 nearly one million smaller cars were sold and 60 percent of those were Volkswagens. The other 40 percent were almost entirely Japanese imports, doubling their sales of 1967. Afraid to lose any market segment, the American auto industry started designing smaller cars again as it had a decade earlier. Ford was ready for the small-car market in 1969 with the midyear release of its Maverick and was working on an even smaller car (Pinto) for a 1970 release. In October 1968, Chevrolet announced it was working on a new smaller car for the 1970 model year. It was code-named XP-887 and would be powered by an all-aluminum engine. The Lordstown, Ohio, assembly plant was being readied for the XP-887, what we would later know as the Vega.

Corvette

The release of the new Corvette in 1968 equaled the excitement of the 1963 Sting Ray. The body was a radical departure from anything ever produced for the street. The 1968 Corvette's roots, as with many Chevrolet vehicles, were founded in a car built for the show circuit, the Mako Shark II.

The Mako Shark II had all the elements Bill Mitchell found attractive in a sporty car. A long nose, short rear deck, highly arched fenders over the

The Caprice Coupe and the Impala Custom Coupe both carried the formal roofline shown here. Available only on the Caprice were hidden headlights that are just barely visible in this photo. This Caprice is also equipped with simulated mag wheel covers first seen in 1965 on the Z-16 Chevelles. *Muscle Car Review*

The Chevelle was completely redesigned for 1968. The new body looked more massive due to a wider tread, but was actually slightly shorter than its predecessor. The 396 Super Sport was one of four distinct Chevelle series. Hidden wipers were standard on all but the 300 series.

wheels, and a boat tail rear deck. Mitchell turned to Larry Shinoda to interpret his ideas on paper, as he had done with the 1963 Corvette. The finished design had a long, pointed nose, no front bumpers, and hidden headlights. The hood had a pronounced dome over the engine. One of the most striking elements of the Mako Shark II was its paint scheme. It was painted in graduated shades of gray, light along the bottom to dark on the upper surfaces. The inspiration for the color came from a Mako shark Mitchell caught while on a fishing trip to Florida. He had the shark mounted and wanted the show car painted the same way. "Inspirations—you've got to keep your eyes open," was Mitchell's comment to Chuck Jordan about the Mako Shark II's paint scheme.

There were two Mako Shark II vehicles: one was a mockup, used only for shows, and the other was what, in the automotive community, is commonly called a "styled driveable." The driveable Mako Shark II was built on a standard Corvette chassis and made its debut at the Paris Auto Show in October 1965. It differed externally from the original in that it did not have the side-mounted exhaust. The driveable Mako Shark II also contained the kinds of gadgets Mitchell loved. The hidden wiper door made it to the production Corvette, but other gimmicks, like the retractable rear bumper, would not.

The design of the production 1968 Corvette actually began in 1965. Duntov and his group of engineers wanted a midengine car for the next gen-

eration Corvette. While they were working on their midengine proposals, Mitchell's group worked on derivatives of the Mako Shark design based on the existing Corvette platform. Duntov's midengine car would have cost millions in tooling costs. *Corvette News,* a magazine published by Chevrolet for Corvette owners, surveyed its readership about their thoughts on the future Corvette and got some interesting results. Well over 95 percent wanted Chevrolet to stay with the front-engine, rear-drive design. By the same margin they let Chevrolet know fiberglass was the material of choice for America's sports car. It was decided to go with the Mitchell-directed Shinoda design on the carryover platform.

Auto design is a tricky business, especially when you must work within the constraints of an existing chassis and engine package. "We are not like other auto companies who take last year's design and add a few wiggles. We wiped the slate clean," said Chuck Jordan. "The trick is to get it to look like a Corvette—a *new* Corvette—not another Corvette." A *new* Corvette is exactly what Larry Shinoda passed from his drawing board in the advanced studio to the production studio where, under the direction of David Holls, Henry Haga's team would refine the design of the new Corvette for production.

The boat-tail design of the Mako Shark II was eliminated, as it restricted the flexibility of building a coupe and a roadster with the same body. The coupe rear window was pulled to a vertical position behind the seats, similar to a Ferrari Dino. This made the coupe upper as short as the convertible and enhanced interchangeability. The rear deck was short with a bobbed spoiler on the back. A pair of Astro ventilation outlets were located on the rear deck just forward of the fuel filler.

The original production design called for a "soft" front bumper similar to Pontiac's Endura-coated bumpers. Bill Mitchell liked the thin, chrome-plated blade bumper design for the front of the production Corvette, and that's what was selected. The dimensions between the bumper and the body were tight, creating a smooth frontal design and a bumper that was not practical. Because of the low seating position and long front overhang, many Corvette owners had a difficult time parking because they couldn't judge the length of the front.

Below the thin front bumper were the grille openings on each side of the front license plate mount. Each grille was made of black plastic and was surrounded by a thin chrome molding. Behind the thin horizontal grille bars was a black plastic modesty panel, which covered the hidden headlights when in the retracted position. The small grille area and resultant headlight hardware blocked normal radiator airflow. To allow air to reach the radiator, engineers opened up the chin area below the grilles and added a small lip just behind those openings to create a high-pressure area to force air up to the radiator. A postproduction fix added a small black plastic spoiler to that lip on the big-block cars to help alleviate their cooling problems.

Two different hoods were used on the 1968 Corvette. The hood profile of the small-block cars was low and had a small, domed center that ran the length of the hood, ending at a peak in the front. Big-block cars had a wide domed area that started close to the front edge of the hood. There were two rear-facing nonfunctional vents recessed near the front with the numbers 427 in chrome on each side.

The long front nosepiece housed the hidden headlights. They were vacuum operated by pulling the light switch to the headlight position, or they could be opened anytime with the override switch at the base of the steering column. The owners manual suggested using the override to keep the lights in the open position during icy weather to prevent the lights from freezing in the down position. If there was a failure in the system, the vacuum hose could be pulled off of the actuator and the head-

This rear view of a 1968 Impala Sport Sedan clearly shows the distinctive quarter panel shape. The Chevy's horseshoe-shaped taillights were set into the rear bumper; the Impala and the Caprice had six taillights, and all other models had four. *GM Media Archives*

light could be lifted manually into the open position. The door concealing the windshield wipers was also vacuum-actuated. When the wipers were turned on, the door rotated up and forward on its hinge linkage. An override next to the headlight override allowed the door to remain in the open position. Both new features added to the 1968 Corvette's mystique. They also added to the complexity of assembly and service.

Etched into the curved and ventless door glass was the "Astro Ventilation" script. The door on the 1968 Corvette was much thicker than its predecessor, since the design was in preparation for federally mandated door guard beams. The door handle on the 1968 Corvette was a new design using standard technology. The push button remained, but the traditional chrome grab handle was replaced by a recessed area covered by a folding chrome door that matched the shape of the door skin.

The new Corvette body had a structural backbone of steel that surrounded the passenger compartment. Large reinforced steel members made up the assembly of lock pillars, hinge pillars, rocker sills, cowl, and on the coupe, a continuation of the hinge pillars over the passenger compartment. This transverse member, resembling a roll bar, was tied to the header bow and provided the T for the T-tops. The original design of the new Corvette coupe called for a one-piece targa-type top. The problem with the targa top configuration was excessive body flex to a point where the doors wouldn't open if the car was parked on a hill. Making the body rigid enough would have added excessive weight to the car. The T-top design and resultant body structure were last-minute changes worked out by engineer Alex Mair and Dave Holls' design team.

The new Nova featured the long hood and short rear deck styling so popular in 1968. The Nova's body was unitized. Its cowl and subframe were shared with the Camaro. *GM Media Archives*

Unlike the Mako Shark II, the production Corvette had rear bumpers. They were similar in shape to the rear bumpers on the 1967 Corvette. Above each bumper section on the flat rear panel were two large round taillights. Between the lights the name CORVETTE was spelled out in small, chrome, block letters.

The beautiful new Coke bottle–shaped Corvette body demanded some concessions in interior space. Shoulder room was considerably reduced by thicker doors and much bulkier door trim panels. There was very little room between the steering wheel and the upper portion of the door trim panel, where a door pull had been concealed under a thick beltline pad. Duntov had been vocal about the beef of the doors until he had an incident on the Milford test track. While on a high-speed pass, something went wrong with the Corvette Duntov was driving and he came to a grinding halt with the driver's door against the guardrail. Duntov escaped uninjured and following the mishap, much less was said about the doors. A midyear change thinned out the trim panel and added a door assist handle similar to the previous Sting Ray's design.

The Impala Sport Coupe carried the same fastback roofline and elliptical wheel openings as the 1967 Impala. Federally mandated side-marker lights were used on all 1968 vehicles. GM designers neatly disguised the front fender-marker light into a chrome bezel that also indicated engine displacement in cubic inches. *GM Media Archives*

The Chevy II was completely redesigned and given a new name for 1968. Now called the Nova, it was offered in only two body styles: a two-door sedan and a four-door sedan. This Palomino Ivory Nova coupe is equipped with the Custom Exterior package, which added bright window moldings and a bright rocker molding with a black accent band. *GM Media Archives*

The new Corvette's interior took a little from the Camaro and a little from an aircraft cockpit. In front of the driver were two large circular housings for the tach and speedometer. The center of the instrument panel housed four gauges and a clock with the radio below. This instrument section swept down to a wide and high center console. It housed climate control, shifter, and ashtray, along with another Corvette innovation, fiber-optic light monitors. Attached to each of the Corvette's primary exterior lights was a piece of fiber-optic cable. It was routed to the console to give the driver an indication that a light may have burned out. Like most of the gadgets on the new Corvette, fiber optics added nothing to the car's performance, but added highly to the "gee-whiz" factor. On the plastic console splitting the seats was the handle for the parking brake. Behind the seats was the small luggage compartment. Access was limited, since one had to fold the front seats forward and angle the items being stowed into place. With the convertible top down, anything much larger than a briefcase wouldn't fit. At the forward edge of the luggage compartment were three doors, two of which could be used for stowing small items, while the third, directly behind the driver's seat, was the battery compartment.

The engines for the 1968 Corvette were the same as those offered in 1967. The only changes were lower-profile intake manifolds and a Rochester Quadrajet carburetor to replace the Holley on all but the highest horsepower engines. The antiquated Powerglide was replaced by the Turbo-Hydra-matic

400 as the only automatic transmission available for the 1968 Corvette. The frame and basic chassis were carried over from the 1967 Corvette, with changes made to the front and rear of the frame for radiator and bumper mountings. The wheel rims were widened to 7 inches to accommodate the standard F70x15 wide-oval tires. These wheels and tires increased cornering ability, but adversely affected ride quality, a trade-off Duntov gladly accepted.

There were no outstanding engineering or performance features on the 1968 Corvette for Duntov to brag about, although it did have an abundance of slick toys such as hidden wipers and trick door handles. The new Corvettes also had a lot of fit and finish problems as it came from the factory. Body integrity and assembly were not as refined as on the previous Sting Ray. Highly touted Astro ventilation provided inadequate interior cooling for the tight interior, and close tolerances between the wiper system and its door caused problems with the door resting on wipers not fully seated in their down position. The newly designed linkage had some owners complaining of wipers crashing into each other as they crossed the windshield. The new T-top roofs on the coupes leaked, as did the auxiliary hardtops on the convertibles. The hotter thermostats and cramped engine compartments created overheating problems, especially on big-block cars. The new Corvettes were so bad, *Car and Driver* magazine refused to road test the new Corvette that Chevrolet provided. It was returned with a lengthy gripe sheet of minor but irritating problems.

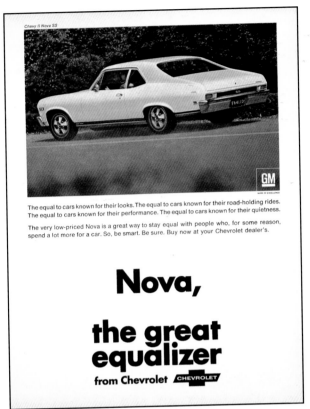

The new 1968 Nova was available in a Super Sport model. It featured a black-accented grille, simulated air intakes on a special hood, red-stripe wide oval tires, and Super Sport emblems. At the beginning of the model year, the only engine available for the Super Sport was the new 350-cubic-inch small-block engine rated at 295 horsepower. Later in the year, the 396-cubic-inch engine was available.

The long front overhang disturbed the Corvette's balance and, to make matters worse, the new body added 100 pounds to the total weight of the car. Under hard acceleration and at high speeds, the front end lifted, causing the car to wander. Stiffer rear springs were added to compensate, even though stiffening the suspension ran contrary to the image of the Corvette as a boulevard cruiser.

Buyers seemed to be ordering the Corvette because of its looks and sexy image, instead of for its performance. This is evidenced by the increase in the addition of options like air conditioning, power steering, and automatic transmissions. Chevy's chief stylist at the time, Dave Holls, was quoted by noted automotive author Karl Ludvigsen in his book *Corvette: America's Star-Spangled Sports Car:* "My boss, Bill Mitchell, who is always very involved with Corvette programs, says you still have to design Corvettes so there is that black one that looks good when it pulls

Chevy's new Corvette coupe was originally intended to have a one-piece targa-type top. The weight penalty incurred by the additional body structure needed to support that style of roof was too great.

up in front of a nightclub." Even with its problems, the beautiful new Corvette sold well, 28,566 units.

Chevy II Nova

Overshadowed by the release of the 1968 Corvette was the all-new Chevy II Nova, formerly referred to as the Chevy II. It was a mom-and-pop car that couldn't compete with the sexiness of the new Corvette. Initially, the Chevy II Nova was ignored at the press previews, yet under its plain exterior was the simplicity and quality engineering for which Chevrolet was historically recognized.

The new Chevy II Nova followed the trend of the 1960s by growing longer and lower. The wheelbase grew by 1 inch to 111.0 inches, and the overall length grew by 4.7 inches. Its height was reduced by 1 inch and both front and rear tread grew by over 2 inches. The 1968 Chevy II Nova came in two body styles, a two-door sedan and a four-door sedan. The styling picked up cues from the new Chevelle with its semi-fastback roofline, and the front-end sheet metal was similar to that of the 1967 Chevy II, featuring peaked fenders and large, single headlamps. Bumpers fore and aft were full face.

Much of the new Nova's engineering was shared with the Camaro. The front suspension and subframe were identical, and the rear suspension was also the same as the Camaro's. The monoleaf spring used on the 1967 Camaro and former Chevy II was replaced with a multileaf spring for all optional engines. Several engines were available, from the 90-horsepower four-cylinder to the 375-horsepower big-block.

Once journalists realized the Corvette was a lot of sizzle with very little steak, they took a more serious look at the Nova. *Hot Rod* magazine called the Chevy II Nova the "sleeper of the year." When *Car and Driver* did its road test of a 375-horsepower 1968 Chevy II Nova SS, it echoed those same sentiments. "The sleeper appeals only to the most secure and sophisticated performance car fancier. There

Bill Mitchell

Bill Mitchell had big shoes to fill in 1958 when he was promoted to vice president in charge of the General Motors design staff. He followed the legendary Harley Earl into one of the most influential positions in the industrialized world.

Bill Mitchell was born on July 2, 1912, in Cleveland, Ohio. He loved cars and at an early age developed a remarkable talent for sketching them. Young Mitchell often made illustrations of the many cars his father drove in his business as an auto dealer. Mitchell studied at several East Coast art schools prior to joining the Baron Collier advertising agency, where he prepared layouts and advertising illustrations. The three brothers who owned the agency were interested in auto racing and founded the Automobile Racing Club of America (ARCA). In 1931, Mitchell became the official illustrator of ARCA and designed its distinctive badge.

In 1935, a friend of Harley Earl, then head of the General Motors Art and Colour Section, suggested that Mitchell send some of his drawings to Earl, who was always looking for talented designers. Mitchell's portfolio won him a position working on what would become the 1937 Cadillac. At the tender age of 24, Mitchell was promoted to chief designer for the Cadillac Studio. In January 1942, he was commissioned into the U.S. Navy as a lieutenant junior grade. He worked in the naval training command, illustrating and laying out flight training manuals. Following the war, he resumed his position as director of the Cadillac studio.

On May 1, 1954, Mitchell was named director of styling under vice president Harley Earl. In that position he served a four-year apprenticeship, which led to the role as the most powerful designer of the 1960s: vice president of styling for General Motors. His late-1950s design of the Sting Ray Racer was a harbinger of things to come from the Mitchell era. That design evolved into the classic lines of the 1963 Sting Ray. Other 1960s-era GM vehicles to receive the distinctive Mitchell touch were the 1962 Pontiac Grand Prix, 1968 Corvette, 1967 Camaro, 1963 and 1966 Buick Riviera, 1966 Olds Toronado, 1965 Corvair, and 1967 Cadillac Eldorado—all outstanding examples of automotive design.

Mitchell had many of the same personality traits as Harley Earl. He ruled through force and fear and, at times, had an awful temper. Once, during a fit of anger in one of the studios, he kicked a bumper and broke his foot. When he entered that studio the next day, no one asked about the plaster cast on his foot.

He was also feisty. During a Cadillac design review with Jack Gordon, GM president at the time, Mitchell's

Bill Mitchell and the Mako Shark II.

Cadillac fins were criticized for being too high. Mitchell liked the fins the way they were and thought Gordon was a fuddy-duddy. To retain his design, Mitchell instructed the clay modelers to increase the height of only one of the clay model's fins for the next design review. During that review, Mitchell, while pointing at the original fin height, told Gordon how he had lowered the fin per his request. Gordon, looking at the other side of the car with its higher fin, approved the design.

What's good for the goose was also good for the gander. Mitchell was dissatisfied with the paint blend on the Mako Shark II, as he wanted the paint to blend from white on the bottom to dark blue on the top, just like the actual Mako shark he caught and had mounted in his office. Every time the paint shop painted the car, Mitchell brought out the mounted shark and compared the colors and blend. Never quite satisfied, he had them repaint the car again and again. One night after Mitchell went home, the boys in the paint shop repainted his shark to match the car. The next day he looked at the car and said, "Now you've got it, guys. That's perfect."

Bill Mitchell could generate enthusiasm by the force of his personality. When working on a car that he had a particular passion for, such as the Corvette, he had a special twinkle in his eye. He was the true architect of the 1963 and 1968 Corvettes. When a young stylist crossed the line and offered too many suggestions on how the Corvette should look, Mitchell would say, "Don't flatter yourself kid; I do the Corvettes around here."

Mitchell once said regarding car design, "It's immoral to make an ugly car. The cost of designing, engineering, and manufacturing are virtually the same—ugly or beautiful." Bill Mitchell *knew* what a car should look like.

are no admiring glances from onlookers to bolster the ego. The entire driver satisfaction is based on the inward confidence that you can put the hurt on a strutting GTO or Mopar before they even realize you're a threat."

The Chevy II Nova had the same general look, feel, and personality as the 1955 Chevy. It was a bit boxy, but simply and elegantly engineered. It offered the consumer a low-cost economy car or a high-performance sedan with better performance stats than a comparably equipped Chevelle SS. It bridged the gap between family car and hot rod. *Car Life* magazine was so impressed with the new Chevy II Nova that it was named one of the magazine's 10 best cars for 1968. "The Chevy II Nova SS presents a sensible, not terribly expensive package that comes nearer to satisfying all the people all the time than anything else we've tested this year. It's a car that's extremely eager to please, and for all around driving anywhere, it has enough personality combined with looks, handling, stamina, and economy to fit nearly any American's image of himself." Consumers loved the Chevy II Nova and bought more than 200,000 in 1968, more than doubling the number sold in 1967.

Chevrolet

The 1968 full-size Chevys received a major face-lift in the form of new front sheet metal and bumpers. The new front end retained the traditional egg-crate grille design, but the bumper reverted to the blade design first shown on the 1965 model. Instead of wrapping around the ends, it terminated at the fenders with a large vertical section. As in 1965 and 1966, the grille was continued below the bumper and had a lower chin panel painted silver. Available as an option on the Caprice were vacuum-operated hidden headlights which, when closed, gave the front a very wide, massive look.

Hidden wipers were standard on all full-size Chevys. To conceal the wipers, Chevrolet designers extended and raised the rear edge of the hood, eliminating the standard cowl vent panel. Near the rear edge of the hood was a long row of slots that allowed air to flow to the plenum. This was a simple solution to the problem of concealing an unsightly, but necessary, set of wipers.

The rear of the full-size Chevy was updated by adding a new bumper with a thin body color valence panel below. The taillights were mounted in the bumper and the shape almost recaptured the circular shape of the taillights by which full-size Chevrolets had been identified for several years. The new lights were more horseshoe-shaped, flat across the top.

Chevy new car sales were so good in 1968 that the company placed this full-page ad in the May edition of *Hot Rod* magazine to help clear out trade-ins. Knowing that performance engines were hot sellers, the ad began with engine numbers (350, 396, and 427) that would get the attention of enthusiasts.

Continuing the trend of upgrading the full-size product, Chevrolet introduced the Impala Custom Coupe in 1968. The Impala Custom Coupe was an Impala two-door hardtop with the formal coupe roofline of the Caprice replacing the fastback top. Appropriate interior upgrades were added to round out the package.

The Caprice was available in three models: a coupe, a station wagon, and a four-door hardtop. The Caprice two-door sport coupe featured full ventless door glass with Caprice script in the lower front corner of the glass. Astro ventilation was also standard on the Caprice coupe and optional on the other Caprice models and Impalas.

Astro ventilation was a system that offered balanced front and rear passenger air distribution under all driving conditions. Outside air was taken in through the vented hood into the plenum. Air ducts distributed the air to two ball outlets in the instrument panel and to two kick panel vents. Each outlet had a positive shutoff; the airflow could be controlled independently. Positive airflow from the instrument panel circulated over the front and rear passengers and exited through two, one-way check valves mounted in the rear seatback panel. The air was then routed to a pressure relief valve in the lock pillar, where it was vented to the outside. Since most Caprice owners ordered air conditioning, the full effect of the Astro ventilation system was never realized.

The Super Sport was demoted from its distinct model position in 1967 back to option status in 1968. The 427SS option was also available, but there were few takers. The trend in performance cars was toward smaller vehicles like the Camaro and the Chevelle SS396. Large Chevys, once the staple of transportation for the average driver, were becoming status symbols.

Chevelle

An entirely new Chevelle with a fastback look was introduced in 1968. The new Chevelle was styled with a long hood and short deck. The tread was 1 inch wider and the wheelbase was 3 inches shorter than its 1967 predecessor, but the new Chevelle was only a fraction of an inch shorter in overall length, yet it did looked longer due to its semi-fastback roof styling.

The 1968 Chevelle was available in four series: the base Chevelle 300 in 12 models, the upgraded Malibu in 10 models, the Concours wagon, and the SS396, available only as a two-door coupe or convertible. Hidden wipers were standard on all Chevelle models except for the 300 series. Chevy designers used the extended hood configuration, similar to that of the Caprice, to hide the wipers.

Super Sport Chevelles were classified as distinct models for 1968, and they came in two sporty forms: a two-door hardtop and a convertible. All Super Sport models came with a 396-cubic-inch engine. Blacked-out trim was applied abundantly to the SS models. The grille was almost entirely black, as was the panel between the taillights. On lighter-colored cars the rocker panels were painted black. On cars of darker colors, the rockers were painted the same color as the body. The Super Sport models had contrasting body stripes that wrapped over the hood and down the side just above the rockers. The SS had a twin dome hood with nonfunctional chrome grilles in the rear. Distinctive SS396 badges were placed in the center of the grille and rear deck. The side-marker light bezel on the front fender of the SS carried the numbers 396, and the taillights wrapped around to provide a rear, side-marker light.

The power line-up for 1968 was the same as for 1967: a base 325-horsepower 396-cubic-inch engine

Chevrolet took the time to explain why those who couldn't afford a new 1968 Corvette should buy a Camaro.

or the hotter, 350-horsepower as the optional motor. A 375-horsepower engine was available in limited production as option code L78, an engine not advertised in the sales literature. To get a Chevelle SS with this optional engine meant a long wait for delivery. You also had to "know someone" who could place this special order. Only 4,751 SS Chevelles with the L78 option were sold in 1968.

The 1968 Chevelle was an outstanding car. More than 400,000 Chevelles were sold, and of those 62,785 were SS396 models. The SS gave the performance buyer a wide range of options with a new dash of style. The 396-cubic-inch engine was in its fourth year of production and was on its way to becoming a legend. By 1968, the SS396 Chevelle had become an icon for the term "musclecar."

Camaro

The basic Camaro for 1968 was a carryover of the 1967 model with a few changes for a fresh appearance. The grille was given a much deeper section and its horizontal bars were colored silver, and the parking lights were moved from the front valence panel and placed in the grille opening. Vent windows were removed from all the Camaros, a

Chevy stylists looked back to the 1965 model for inspiration to design the front of the 1968 Chevy. The thin blade bumper returned, splitting the egg-crate grille. The valence panel below the grille was painted silver on all models. This Super Sport convertible is equipped with front disc brakes, as evidenced by Rally wheels. *GM Media Archives*

move touted by Chevrolet as a feature that improved sealing, reduced wind noise, increased visibility, and reduced theft. It also reduced engineering time, as vent windows are complicated to design properly. And a car with full door glass is also much less expensive to produce and assemble.

The Camaro chassis was also carried over with the exception of the rear suspension changes. The radius rods added to the V-8 models in 1967 were removed and the single-leaf spring was replaced with a multileaf heavy-duty unit. The rear shocks were staggered, which helped counter rear axle wind up on hard acceleration.

The year 1968 witnessed the full blossoming of the Z/28 option. First introduced in 1967 with limited availability, the Z/28 was now widely available and highly promoted by Chevrolet's marketing staff. Feedback from teams racing Z/28s in 1967 were responsible for chassis improvements on the 1968 models. Rear disc brakes were also added as an option, a first for Chevrolet in a passenger car. It was the closest thing yet to a Corvette, and Chevrolet proudly proclaimed that fact in its advertising.

Sales for the Camaro's sophomore year were outstanding with 235,151 units sold. Ford's Mustang still outsold it but by a much narrower margin than in 1967.

Corvair

The basic styling of the Corvair was unchanged for the 1968 model. The Corvair line was now down to two body styles, a hardtop and a convertible (the four-door was dropped). Two series of Corvairs were available, a 500 (coupe only) and the upgraded Monza.

In January 1968, *Car Life* magazine ran the last Corvair road test. The editors loved the car and stated it was a true driver's car. Their complaints could be attributed only to poor assembly, not engineering. Corvair sales continued to slide in 1968 to 15,399 units.

Chapter 10

1969 The Decade of Style and Speed Draws to a Close

The average price increase for a new General Motors car in 1969 was only 1.6 percent, far below the actual increases in costs due to higher labor and material charges and government-mandated safety equipment. These higher costs reduced the company's net income in a year of increased sales. General Motors sold 5.26 million units in 1969, which accounted for 52.4 percent of the domestic auto market. Chevrolet's six models sold a total of 2.1 million cars, once again outselling its perennial rival, Ford.

Emissions systems on 1969 Chevrolets were carried over from the 1968 models with very little change. The problems General Motors was having with auto emissions systems took place in the courts, not in the factory. At the request of California authorities, GM in 1952 joined other members of the motor vehicle industry in launching a program of cooperative research and development to resolve the problems of motor vehicle air pollution. Starting in 1955, the auto industry had a cross-licensing agreement, which made possible the free exchange of technical information and patent rights. This program was a matter of public knowledge and was successful in reducing hydrocarbon emissions to a level 63 percent below that of pre-controlled models. Yet on January 10, 1969, the

The 1969 Chevelle was essentially a carryover of the 1968 model with the exception of a new grille and rear quarter caps. The SS396 was no longer a separate model, but was an option to the Malibu coupe, convertible, or 300 Deluxe model sedan. The SS featured 14x7-inch five-spoke styled steel wheels with a bright trim ring. *GM Media Archives*

The 1969 Camaro received a minor face-lift on the lower body for 1969. The overall appearance was a wider and lower car. This 1969 Z/28 also has the Rally Sport option featuring hidden headlights. The hood is the Shinoda-penned cowl-induction version, an option on the Z/28, SS350, or any Camaro with a 396-cubic-inch engine.

Department of Justice filed a civil antitrust suit in the United States District Court in Los Angeles, naming as defendants General Motors, Ford, Chrysler, American Motors, and the American Manufacturers Association. The complaint sought injunctive relief with respect to the industry's cooperative effort to develop emission controls.

On October 29, 1969, a consent judgment terminated the suit. The final judgment imposed limitations on the future exchange of technology in the emissions field. A GM press release regarding the judgment said: "General Motors has accepted this settlement of the Justice Department suit in the belief that the interests of the Corporation and the public will be better served by our getting on with the job of solving the emission control problem rather than engaging in a long antitrust trial." This judgment hurt the smaller auto companies that didn't have the vast research and development resources of GM. Technology developed by GM, like the PCV valve and air injection reactor, had been shared with auto producers large and small. Now each manufacturer was on its own to solve the same problems.

New standard safety equipment for 1969 Chevrolets included headrests for the driver and

Chevrolet engineers had tinkered with 302-cubic-inch versions of the small-block as early as 1958. The Z/28's 290-horsepower 302 was a combination of the 4-inch bore of the 327 block and the 283's 3-inch-stroke crankshaft. A free-breathing intake, big valves, and a long-duration camshaft were added to produce a high-revving powerhouse.

This Biscayne two-door sedan was the least expensive full-size Chevy body style that could be purchased in 1969. This owner, instead of wasting money on fancy trim, put his money under the hood. This sedan is powered by a 427-cubic-inch engine. *Donald Farr/Muscle Car Review*

One option missing from the 1968 Corvette was the side-mounted exhaust. It returned in 1969 with a stylish set of chrome-plated covers. Fitted to all 1969 Corvettes were 8-inch-wide rally wheels (1968 rally wheels were 7 inches wide) with trim rings and a center cap. This particular Corvette is also equipped with an optional factory installed alarm system. The lock to activate the system is located on the rear panel above the Corvette letters. *Corvette Fever*

front seat passenger on all Chevy models. Headrests were first seen on Chevys as options in 1966 and were successful in reducing the effects of whiplash most commonly associated with rear-end accidents. Side impact bars were installed on all full-size Chevrolets and consisted of a welded steel, two-piece construction forming a box section that extended from the lock pillar to the hinge pillar within the door. They were designed to reduce passenger injury in the event of a side impact.

A "skid-plate" was hidden under the headliner of Camaro, Chevelle, Chevy II Nova, and full-size Chevy hardtop models at the windshield header. This plate welded to the header structure was designed to provide a sloping, edge-free surface to absorb occupant head impact in an accident. Automatic-locking front seatbelt retractors were standard on Chevrolet and Chevelle models, and Corvettes featured a new inertia shoulder belt retractor. These latter units allowed the occupant to move forward slowly against the belt,

A 1969 Nova two-door with a 307-cubic-inch engine, such as the one shown here, listed for $2,405. Sales figures for the Nova dropped to about half of what was sold in 1968. *GM Media Archives*

but they locked up when forward motion was rapid, as in the case of an accident.

In an attempt to reduce vehicle thefts, Chevrolet changed the ignition lock on all models except the Corvair. The new ignition switch was moved to the steering column, and when the key was removed, it locked the ignition system, steering wheel, and transmission. The key could only be removed when it was in the LOCK position, and with the key in the OFF position, the transmission lever and steering wheel could be moved. Two wing tabs aligned with the new square ignition key to provide leverage when turning the ignition lock. The LOCK position was operative only when a car equipped with an automatic transmission was in PARK, or when a car equipped with a manual transmission was in reverse. Two separate keys were used on all 1969 Chevys. The square key was used for the ignition and door; the oval key was used for the trunk, glovebox, and console. Another antitheft feature on all 1969 Chevys (except Corvair) was the relocation of the interior door lock knobs. Depending on the model, the lock knobs were moved forward either 3 or 12 inches. The new location made access easier for driver and passenger and created an obstacle for anyone trying to unlock the door from the outside with a coat hanger.

The 350-cubic-inch engine, first introduced in 1968, replaced the 327-cubic-inch engine in heavy-duty applications. In 1969, the 327 was available only with a two-barrel carburetor rated at 210 horsepower for the Camaro and rated at 235 horsepower for the full-size Chevy. All 302, 327, and 350-cubic-inch cylinder blocks were redesigned, strengthening the lower block walls and main bearing structure. Strengthened blocks provided increased durability by producing a smoother transfer of the forces created by combustion and reciprocating components. Attaching bolts were lengthened for all bearing caps. The longer bolts spread the load pattern deeper into the bulkhead, reducing parting line stresses. Minor changes were made to the small-block cylinder heads for accessory attachment.

Chevrolet announced the release of the Turbo-Hydra-matic 350 transmission in 1969. It was a fully automatic three-speed transmission enclosed in a two-piece aluminum case and was available as an optional automatic transmission for all six-cylinders and small-block V-8s. It was similar to the Turbo 400 in design, but with a lighter-duty construction. The

Vince Piggins

Vince Piggins was affectionately known as Chevrolet's "Mr. Performance." He joined Chevrolet in May 1956 to coordinate the exposure of Chevrolet products, including the new small-block V-8 engine, in racing activities. He was a tall, soft-spoken man who worked directly with all breeds of Chevrolet racers. He spoke the racer's language and knew how to maneuver within Chevrolet's corporate structure. Every weekend Piggins attended some form of racing, from hydroplanes to sprint cars. Every professional racer who ran a Chevy engine was on a first-name basis with him, and he was the direct conduit between the racer and the factory.

The Z/28 Camaro was the brainchild of Piggins. Prior to the release of the first Camaro in 1967, Piggins worked both ends of what turned out to be a sweet deal. In 1966, the SCCA (Sports Car Club of America) was developing a production sedan class, and cars running in this class were production-based sedans with a 305-cubic-inch limit. Piggins convinced SCCA officials that with Chevrolet's support, this new class could be a success, and he suggested to General Manager Pete Estes how easy it would be to put a 283 crankshaft (with a 3-inch stroke) into a 327 block (with a 4-inch bore) and have a 302-cubic-inch engine. Estes, a former engineer, bought into the concept immediately, and the Z/28 was born.

The Z/28 option was available soon after the release of the 1967 Camaro. The Z/28's high-performance 302-cubic-inch engine was conservatively rated at 290 horsepower at 5,800 rpm and could easily rev to over 7,000 rpm.

The Z/28 was a brilliantly conceived and executed vehicle. It offered the young family man Corvette performance at half the price. Its winning ways attracted

Vince Piggins

buyers to Chevrolet dealerships with money in hand. They might not have bought a Z/28, but they saw Chevrolet's vision of speed and style—a vision created in a large part by Vince Piggins.

Turbo 350's internal components varied slightly with different engine combinations. There were two torque converters, one for all engines up to 350 cubic inches and one for all 350-cubic-inch engines. Several clutch pack combinations were used depending on the engine to which the transmission was attached.

A new second-generation disc brake design was introduced for the Camaro, Chevelle, Chevy II Nova, and full-size Chevy. This new design featured a single piston with a floating caliper. It replaced the previous fixed-caliper four-piston design. The new disc system included a new dual-circuit master cylinder with larger reservoirs to handle the increased fluid volume. New brake rotors were also specified. The full-size Chevy was equipped with 11.75-inch diameter by 1.25-inch thick rotors, while the other car lines had 11x1-inch rotors. A 15x6-inch wheel rim of conventional construction was specified for the full-size Chevy with disc brakes, and rally wheels were an option.

The Chevy II Nova, Camaro, and Chevelle used 14-inch wheel rims with their disc brakes. Disc brakes were a required option on the SS396 Chevelle, Chevrolet SS427, Chevy II Nova SS with a 350 or 396 engine, Camaro with a 350 or 396 engine, and Z/28. Standard drum brakes for the Chevelle, Chevy II Nova, and Camaro were upgraded with

The 1969 Caprice coupe shared its unique roofline and concave rear window with the Impala Custom Coupe. The Caprice could be ordered with hidden headlights, which added to the wide appearance of the Chevy's front end. *GM Media Archives*

finned drums and heavy-duty linings.

Introduced on the 1969 Caprice, Impala, and on all Camaros was a new variable-ratio power steering system. This new system offered the driver faster and more responsive steering during turns, with fewer turns lock-to-lock. More important, it offered the driver-enhanced "road feel" in the straight-ahead position. The variable ratio was accomplished by having gear teeth of different lengths on the pitman shaft sector. The variable-ratio system required a new power steering pump that operated at a higher pressure.

All Chevrolet passenger cars in 1969 were equipped with a tamper-resistant odometer. It featured an antireverse drive that prevented the speedometer from being run in reverse to lower the mileage reading.

Warranties on all Chevrolet passenger cars were reduced to 12 months or 12,000 miles. The extended warranty for the engine and driveline was kept at five years or 50,000 miles. Quality control problems were plaguing all car lines and, within a few years, a massive recall was made for defective motor mounts.

Chevrolet

The full-size Chevrolet was the only car in Chevrolet's 1969 line-up that looked significantly different from its 1968 counterpart. It was based on the same platform as the 1968 and carried two distinct rooflines on two-door sport coupe models. The Impala had a semi-fastback roof with much less slope than on the 1968. The Impala Custom Coupe and Caprice had a roofline that was more upright with a unique concave rear window and streamers which extended the C-pillar along the top of the quarter panel. The new Chevy's front end was aggressively restyled with a halo bumper surrounding the grille. Its design was reminiscent of the 1968 Lemans/GTO without the vertical center bar. By

sheer coincidence, this front-end design appeared on the Chrysler 300 in 1969.

Within the halo front bumper was a new fine-mesh egg-crate grille that was suggestive of a Cadillac grille. The new grille assembly was recessed into the bumper and was constructed of chrome-plated plastic, which appeared richer than the previous anodized aluminum grilles. Different front-end treatments were created by using argent paint on certain sections of the grille on different models. Hidden headlights were optional on the Caprice and Kingswood wagons. The two-door-per-side eyelid design used on the 1968 models was revised to a single door per side that rotated up above the light.

The rear-end design was similar to that of the 1968, featuring a full-width bumper and lower valence panel. The taillights, now rectangular in shape, were placed in the bumper (similar to the 1968 design).

The bodysides started out as a rather simple design, which blended well with the front and rear design. The wheel opening cut lines were the same as those on the 1968 in order to retain the inner wheelhouses. Subtle blisters were added to the surfaces around the front and rear wheel openings. "We thought the sides should have more interest or detailing," said Chevrolet stylist Dave Holls. "I was very interested in European rally cars that started not to have wheel lips, but the surface around the wheel opening faired out. The best execution of this design actually came about on the '70 Chevelle Super Sport with its bigger tires."

One feature that appeared for the first time on a Chevrolet product in 1969 was the electric rear window defogger. It featured silent uniform rear window defogging. The electric rear window defogger was available only on the Caprice and Impala Custom Coupe. The inside of the rear window was lined with a series of 0.03-inch-wide conductive ceramic strips spaced 1.12 inches apart. When the instrument panel ON-OFF switch was set to the ON position, up to 20 amps flowed through the window's ceramic heating element. The heat generated burned off any condensation on the inside and warmed the glass sufficiently to melt snow or ice that may have accumulated on the outside. An electric timer limited the 20-amp flow to 15 minutes and then automatically reduced the amperage to a constant 5 amps. To handle the additional current load, a 63-amp alternator was standard when the electric rear window defogger option was selected.

Working at Chevrolet also meant driving a Chevrolet. When designer Dave Holls moved from Buick (where he drove a Riviera) to Chevrolet, his neighbors thought something was wrong at work, since he was now driving a less-expensive car. "Chevrolet was *the* job," Holls said, "my neighbors didn't know that." Shortly after John De Lorean took over in 1969 as Chevrolet's general manager, he was asked to stop riding in his Cadillac limo and get a Chevy. De Lorean approached Holls and said, "Don't make a lot of noise, but let's design a beautiful Chevy limo using the Caprice coupe upper." Holls admitted that Bill Mitchell was not pleased with the Chevrolet limousine he was instructed to build for De Lorean, who claimed the Caprice limo was not for him personally, but was a prototype for a presidential limo similar to the one Ford Motor Company was building for the president of the United States.

Corvette

Very few perceptible changes were made to the 1969 Corvette. Most changes were small ones on the engineering and assembly side to improve product quality. Clearly absent from the 1968 Corvette was the Stingray identification. Chevrolet designers had always felt the 1963 through 1967 Corvettes were the true Sting Rays. The 1968 was what they considered a new design, so the script was left off. In 1969, the Stingray script, now one word, reappeared above the front fender louvers.

The exterior door handles on the Corvette were changed in 1969 by removing the push button and reworking the press flap grip to actuate the door latch release. The horizontal grille bars that were silver on the 1968 were now black. Back-up lights were moved from the rear valence panel up to the inboard taillight position. The only other major exterior change was the widening of the rally wheels by 1 inch to 8 inches.

New, thinner door trim panels offered the driver and passenger more shoulder room, and a little more interior room was gained with a 1-inch reduction in the diameter of the steering wheel. Map pockets were sewn into the face of the instrument panel on the passenger side, and a small plate was added to the console to indicate the engine's horsepower rating and cubic-inch displacement.

The engine horsepower ratings stayed the same on the 1969 Corvette. The only change was to the displacement of the two small-block engines (300 and 350 horsepower), now 350 cubic inches instead of 327. Transmission and rear-end combinations were also the same. Returning as an option after a one-year absence was the side-mounted exhaust option. The muffler was redesigned from the previous straight-through chambered pipe to a reverse-flow design terminating with a square tip. The side-

On May 14, 1969, the last Corvair was produced, ending 1969 production at 6,000 cars. The only changes made between the 1968 and 1969 models were those required by law. *GM Media Archives*

Several small changes were made to the exterior of the 1969 Corvette. Most noticeable is the addition of the Stingray (now one word) script on the front fender. The press flap grip at the top of the door was modified to actuate the door latch release, eliminating the push button. The horizontal grille bars, painted silver on the 1968s, were now black. In the rear, the back-up lights were moved from the lower valence panel up to the two inboard taillight positions. This particular Corvette coupe is powered by one of the optional 427-cubic-inch engines. This unique hood was found on all Corvettes equipped with a 427-cubic-inch engine. *GM Media Archives*

mount exhaust covers were new and constructed of three, chrome-plated diecastings attached to a Fiberglas frame. The new pipes blended beautifully with the Corvette's body and were almost as loud as the previous design.

Corvette sales for the 1969 model year were 38,762 units, up significantly from 1968. A prolonged sales period that ran into the 1970 calendar year accounted for the large increase in sales.

All full-size Chevys in 1969 had an impact beam within the door. It was a welded structure placed there to reduce passenger injuries in the event of a side impact to the door. This Impala convertible also has headrests. First seen as an option in 1966, headrests were now required by federal law. *GM Media Archives*

Chevelle

The 1969 Chevelle was essentially a carryover of the 1968 model. Full side glass and Astro Ventilation were added to all sport coupe and convertible models. A new grille and rear quarter caps were the extent of exterior changes. Because of the new end caps, a rear side-marker light was added. New 14x7-inch styled steel wheels with a slender five-spoke design and bright trim ring were the only wheels available on the SS396, which was no longer a separate model but was an option in 1969. It could be added to the Malibu coupe, convertible, or 300 Deluxe model sedan.

Chevelles sold extremely well in 1969 with a total of slightly more than 500,000 units, bested only by the full-size Ford and Chevy.

Chevy Nova

There were very few changes to the 1969 Chevy Nova. Body side nameplates were newly styled and located on the front fender. Super Sport options and performance engines were featured in ads in the enthusiast magazines. It was the little car that offered something for everyone at a very reasonable price.

Camaro

Small but significant styling changes were introduced on the 1969 Camaro. The 1967 and 1968 models had been cleaner and smaller-looking. The 1969 Camaro grew in bulk, most noticeably by virtue of its larger grille opening. The slender front bumper design now curled up on the ends, wrapping up into the fender cap, creating a clean, rectangular design. Within the cavernous opening was the large-grid egg-crate grille, which joined at the center in a V shape. The Rally Sport option offered hidden headlights with three horizontal windows in each headlight door, and the taillights were similar to the previous design but were longer and thinner.

The side treatment of the 1969 Camaro was exceptional. The wheel openings went from the circular design with flared lips at the top (used on the 1967 and 1968) to a design where the top edge was flattened off on distinctively beefy fenders. The flats at the tops of the wheel openings coincided with the peak of the crown on the door. Sweeping back from the top edge of the front wheel opening was a horizontal character line that ran rearward on the

The Impala Sport Coupe featured a roofline halfway between the fastback roof of the 1968 and the more upright Caprice roof. All Chevy products in 1969, except for the Corvair, were equipped with a steering column ignition lock. This was added to prevent vehicle thefts. *GM Media Archives*

fender and washed out middoor into the door's crown. The rear wheel opening had the same treatment with the character line washing out at the very end of the quarter panel.

A new cowl-induction hood was optional on the 1969 Camaro, and its design was prompted by Vince Piggins for Z/28s competing in SCCA events. Several proposals were considered, but the one chosen was sketched by Larry Shinoda and featured a simple raised rectangular area in the center sweeping back to the cowl. The rear was opened, allowing high-pressure air at the base of the cowl to be ducted to the carburetor. An electric solenoid was actuated at 90-percent throttle, opening an air valve allowing cool outside air to rush in. The option number was ZL2 and it was available for an addi-

tional $79 on the SS350, Z/28, and on Camaros with a 396 engine. Over 10,000 new Camaros were ordered in 1969 with the cowl-induction hood.

The Camaro and Chevy II Nova shared the same instrument panel in 1969. It was the same as the one first seen in the 1968 Nova, and its appearance was quite different from the Chevy II Nova's due to the unique instrument cluster assembly.

Sales were very good for the Camaro in 1969 with a total of 243,095 units sold. It must be noted that like the Corvette, sales of the 1969 Camaro were extended into the 1970 model year. One of the hottest-selling Camaros was the Z/28, with 19,014 sold. Due to the extended sales year, it was several years before sales of the Z/28 option eclipsed the mark set in 1969.

The side of the 1969 Chevrolet was enhanced with the addition of subtle blisters to the surfaces around the front and rear wheel openings. This design was taken from European rally cars where, instead of flaring the wheel opening, they would have the surface pulled out. *GM Media Archives*

Corvair

The last Corvair rolled down the assembly line on May 14, 1969. It was a coupe and it ended 1969's production of only 6,000 Corvairs. The only changes made to the Corvair in 1969 were those required by law. It didn't even get the locking steering column that all other 1969 Chevrolet models received.

Over its 10-year life span, 1,786,243 Corvairs were sold. They were loved by automotive enthusiasts for their innovative design and sporty look and feel. The Corvair blossomed with the advent of the Monza option and its bucket seats. It was on its way to becoming a world-class sports car when lawsuits and negative publicity forever branded the Corvair as an ill-handling, dangerous car.

In an effort to give each customer exactly the car he wanted, each Chevrolet model had a lengthy list of unique options and requirements. For example, the full-size Chevrolet had two different power steering systems dependent on model, 35 different front springs selected by model and options, and four different drive shafts dependent on model, engine, and transmission selection. This multitude of parts was expensive to design, manufacture, ship, warehouse, assemble, and inventory. The complexity of this system added cost to each car and consumers weren't getting a whole lot more for their money. This monster was in the process of consuming itself in 1969.

When John De Lorean took over from Pete Estes as Chevrolet's general manager in early 1969, he looked at the entire operation and decided some streamlining was needed. Each Chevrolet model was evaluated to see where cuts could be made without affecting the product. Most of these changes would come about in the early 1970s, making the division more efficient and thereby more profitable.

Product quality was another issue De Lorean would face in the coming years. Mistakes of the past would haunt him over his tenure as Chevrolet's general manager. Product recalls for broken motor mounts, exhaust gases leaking into the passenger compartment, and premature rusting were only a few of the challenges ahead for De Lorean and Chevrolet in the 1970s.

Appendix

Bibliography
Articles

"1962–1967 Chevrolet Chevy II: Plucking Falcon's Feathers." *Collectible Automobile* (June 1993).

"1964–67 Chevelle: The 1955 Chevrolet Reinvented." *Collectible Automobile* (August 1992).

"1965–66 Chevrolet: When Bigger Was Better." *Collectible Automobile* (December 1993).

"1967–69 Camaro: Considered Response." *Collectible Automobile* (October 1990).

"360 HP For Chevy II." *Hot Rod* (March 1962).

"A Pre-Retirement Chat with Chuck Jordan." *Collectible Automobile* (December 1992).

"Autos—The New Generation." *Time* (October 5, 1959).

"Big 'Vette." *Hot Rod* (January 1962).

"Camaro: New Kid on the Block." *Hot Rod* (January 1967).

"Can Anyone Build Really Safe Cars?" *Motor Trend* (April 1967).

"Caprice 396." *Motor Trend* (June 1965).

"Car Life Engineering Excellence Award." *Car Life* (February 1962).

"Car of the Year." *Motor Trend* (April 1960).

"Chevelle 396." *Car Life* (September 1965).

"Chevrolet Impala Super Sport 409 V-8 with Powerglide." *Car Life* (March 1963).

"Chevrolet's Mystery 427 V-8." *Hot Rod* (May 1963).

"Chevy II 327/350 V-8." *Car Life* (May 1966).

"Chevy II Nova 327 Road Test." *Motor Trend* (July 1966).

"Chevy II Nova SS Road Test." *Car and Driver* (August 1968).

"Chevy II with a V-8." *Motor Trend* (March 1962).

"Chevy II." *Car Life* (February 1962).

"Chevy's New Flying Wedge." *Hot Rod* (March 1965).

"Corvair Corsa Road Research Report." *Car and Driver* (October 1964).

"Corvair Monza 4-door Road Test." *Motor Life* (May 1961).

"Corvair Monza Spyder." *Car Life* (April 1962).

"Corvette—Hottest Thing on the Road." *Hot Rod* (June 1961).

"Engineering the Chevy II." *Car Life* (February 1962).

"Fords Sweep Daytona 500." *Hot Rod* (May 1963).

"Hottest 'Vette Yet." *Hot Rod* (May 1967).

"How Old a Car Should You Buy?" *Motor Trend* (June 1965).

"Living with Late Greats." *Late Great Chevys* (June 1991).

"Martyr." Special Interest Authors (May–June 1974).

"Motor Trends." *Motor Trend* (June 1961).

"Requiem for a Lightweight." *Sports Car Graphic* (August 1969).

"Spotlight On Detroit–'67 Cars." *Motor Trend* (July 1966).

"Spotlight on Detroit–'67 Cars." *Motor Trend* (July 1966).

"Spotlight on Detroit." *Motor Trend* (December 1962).

"Sting Ray Corvette for '63." *Hot Rod* (October 1962).

"Super Street Chevelle." *Hot Rod* (February 1966).

"Tech Analysis of the '68s—Chevy." *Hot Rod* (October 1967).

"The Big Go West." *Hot Rod* (May 1961).

"The Warranty War." *Motor Trend* (June 1965).

"The Warranty War." *Motor Trend* (June 1965).

"Tres Chevelles." *Motor Trend* (July 1967).

"Whatever Happened To Fuel Injection?" *Car Life* (February 1966).

"Zora Arkus-Duntov." *Hot Rod* (September 1967).

General Motors Corporation. *GM Annual Report* (1960).

General Motors Corporation. *GM Annual Report* (1961).

General Motors Corporation. *GM Annual Report* (1962).

General Motors Corporation. *GM Annual Report* (1963).

General Motors Corporation. *GM Annual Report* (1964).

General Motors Corporation. *GM Annual Report* (1965).

General Motors Corporation. *GM Annual Report* (1966).

General Motors Corporation. *GM Annual Report* (1967).

General Motors Corporation. *GM Annual Report* (1968).

General Motors Corporation. *GM Annual Report* (1969).

Books

Adams, Noland. *The Complete Corvette Restoration & Technical Guide—Vol. 2, 1963 through 1967.* Automotive Quarterly, Inc., 1987.

Bayley, Stephen. *Harley Earl and the Dream Machine.* New York: Alfred A. Knopf, Inc., 1983.

Boyce, Terry V. *Chevy Super Sports 1961–1976.* Osceola: Motorbooks International, 1981.

Colvin, Alan L., *Chevrolet by the Numbers 1965–'69.* Cambridge, MA: Robert Bentley, Inc., 1994.

Colvin, Alan L. *Chevrolet by the Numbers 1960–'64.* Cambridge, MA: Robert Bentley, Inc., 1996.

Craft, Dr. John. *Vintage & Historic Stock Cars.* Osceola: Motorbooks International, 1994.

Cummings, Doris and David, Jr. *Chevrolet Book of Numbers Volume 2 1953–1964.* Blairsville, PA: Crank'en Hope Publications, 1989.

Editors of *Consumers Guide. Cars of the '60s.* New York: Publications International, 1979.

Finch, Christopher. *Highways to Heaven.* New York: Harper Collins, 1992.

Gunnell, John. *Illustrated Chevrolet Buyers Guide.* Osceola: Motorbooks International, 1989.

Halberstam, David. *The Reckoning.* New York: William Avon, 1986.

Herd, Paul A. *Chevelle SS Restoration Guide.* Osceola: Motorbooks International, 1992.

Herd, Paul A. *Chevrolet Parts Interchange Guide.* Osceola: Motorbooks International, 1995.

Hooper, John R. *Big Book of Camaro Data 1967–1973.* Osceola: Motorbooks International, 1995.

Koblenz, Jay *Corvette America's Sports Car* Skokie, Il: Publications International, 1984.

Kimes, Beverly Rae and Robert L. Ackerson. *Chevrolet a History From 1911.* Automobile Quarterly, Inc., 1986.

Ludvigsen, Karl *Corvette America's Star Spangled Sports Car.* Princeton: Princeton Publishing, 1973.

Mueller, Mike. *Corvette Sting Ray 1963–1967.* Osceola: Motorbooks International, 1994.

Nichols, Richard. *Corvette 1953 to the Present.* [city:] Gallery Books, 1985.

Wright, J. Patrick. *On a Clear Day You Can See General Motors.* Grosse Pointe, MI: Wright Enterprises, 1979.

Index